METHANOL FUEL CELL SYSTEMS

Advancing Towards Commercialization

METHANOL FUEL CELL SYSTEMS

Advancing Towards Commercialization

DAVE EDLUND

PAN STANFORD PUBLISHING

Published by

Pan Stanford Publishing Pte. Ltd.
Penthouse Level, Suntec Tower 3
8 Temasek Boulevard
Singapore 038988

Email: editorial@panstanford.com
Web: www.panstanford.com

British Library Cataloguing-in-Publication Data
A catalogue record for this book is available from the British Library.

Methanol Fuel Cell Systems

ISBN 978-981-4241-98-4 (Hardcover)
ISBN 978-981-4303-14-9 (eBook)

Printed in the USA

To my father and mother, who inspired and nurtured my lifelong love of science; to my wife, who has tirelessly supported and motivated me; and to my children, who have given me purpose.

Contents

Preface

Hydrogen. It is the lightest element, the first in the Periodic Table, the primary constituent of stars. It is the fuel that heats and lights Earth. Hydrogen is all around us. Combined with oxygen and/or carbon, hydrogen is present in water, plant and animal tissue, petroleum, natural gas, plastic, paper, and wood. And yet, as abundant as hydrogen is in chemically combined forms, it is not naturally available as the free element.

This presents a challenge for the commercial success of fuel cell technology. Fuel cells ideally operate on pure hydrogen. But where is the hydrogen to come from, and how is it to be supplied to fuel cell products? There are surprisingly few options. Hydrogen manufactured at large plants* may be compressed and distributed by truck in heavy steel cylinders in a variety of sizes. Or, hydrogen may be liquefied (an extremely energy-intensive process) and distributed in heavy cryogenic tanks. These approaches are economically and logistically disadvantaged because of the cost of transporting compressed hydrogen.

In specialized applications, hydrogen may be generated by electrolysis using off-peak electric power; the hydrogen being stored for later local consumption by fuel cells to make electricity. A less expensive option is to generate hydrogen locally from available hydrocarbon or alcohol fuels using a chemical process generically called reforming. Regardless of the fuel, reforming converts hydrocarbons and alcohols—for that matter any carbon-containing material—into hydrogen and a mixture of carbon dioxide and carbon monoxide.

One legacy of the Industrial Revolution is humankind's reliance on thermal engines that produce mechanical work through the

* These large, centralized hydrogen plants may be based on thermochemical processes (traditional), electrolytic processes (water electrolysis), or thermal processes (solar or nuclear) that split water into hydrogen and oxygen at very high temperatures.

combustion of fuels—initially coal or wood and later liquid hydrocarbons and natural gas. Because of this legacy, fuels as we know them have been highly refined for one purpose: to reliably burn under all intended operational conditions. Thus, commercial fuels are blends of many, perhaps hundreds, of organic compounds and additives such that the vapor pressure, viscosity, lubricity, emissions, etc., of the fuel are controlled regardless of the environmental conditions (hot or cold) or source of the petroleum or gas feedstock. Because of this, commercial fuels are very difficult feedstocks for a chemical reactor that is normally engineered to operate with a specific chemical feedstock, rather than a variable blend. In addition, most fuels are derived from either petroleum or natural gas and consequently contain significant concentrations of organosulfur compounds. At temperatures less than about 700°C, organosulfur compounds will poison known commercial catalysts for fuel reforming.

A little-known fuel that has been used for decades in racing engines is methanol, favored for its high octane rating and the fact that it generates little thermal radiation when it burns (contributing to greatly improved safety in the event of an accident). Methanol is also easily reformed to make hydrogen. This is the focus of this book.

Methanol is arguably an ideal fuel for local reformation to make hydrogen. As a primary chemical building block, it is produced around the world from natural gas and other hydrocarbons or from renewable resources, including garbage, sewage, and biomass. Methanol is a global commodity: it is shipped across oceans in large tankers; on land it is shipped in rail cars and by tanker trucks, and in both plastic and steel barrels.

Hydrogen can be chemically extracted from methanol relatively easily at intermediate temperatures by using either air or water as an oxidant. As with all reformation processes, the product hydrogen is combined with carbon monoxide and carbon dioxide. Indeed, it is the formation of carbon dioxide—a chemical conversion associated with a large release of energy—that drives the overall process resulting in production of hydrogen. If water is the oxidant, one-third of the product hydrogen comes from water, making this the preferred route.

Unlike hydrocarbon fuels, methanol is completely miscible with water, and such mixtures will not freeze even in the coldest climates.

Methanol also disperses quickly (by virtue of water miscibility) if spilled. Methanol is a naturally occurring compound that is produced as a by-product of spoiled fruit and the decay of sewage. Naturally occurring microbes metabolize methanol quickly, and if spilled on the ground or in water, methanol is degraded in a matter of days to a couple weeks.

Despite all the reasons that methanol is a good choice for generating hydrogen, it has one major drawback: methanol is poisonous. Of course, the same is true for gasoline, diesel, and other fuels. Even ethanol is poisonous in sufficient quantity. However, methanol has a notorious history of being used illicitly to make cheap booze, frequently resulting in blindness, and even death, of innocent patrons. This history has resulted in an irrational fear of methanol by many individuals, companies, and governments. Only recently has a more balanced perception of methanol emerged, perhaps due in part to the success of SFC Energy AG's direct-methanol fuel cell products sold for recreational use.

As the title suggests, this book will examine in detail the use of methanol reformers in fuel cell systems. The book is aimed at those with a technical interest in methanol reformers as well as a business interest. Chapters are devoted to a discussion of methanol as a fuel, the chemical and engineering aspects of reforming methanol, and practical approaches to hydrogen purification (since relatively high-purity hydrogen is needed for low-temperature fuel cells). A recurring theme is cost reduction, since the purchase price of methanol reformers (and fuel cells) has historically been a significant barrier to widespread commercial acceptance. The latter half of the book addresses potential markets and applications and also delves into the cost issues in detail. These chapters offer value not only to design engineers but also to marketers and managers who are engaged with prospective customers for fuel cell systems and reformers.

Bridging the gap between what a customer thinks he or she wants, and what can be economically delivered, is as much an art as it is a science. Through an understanding of the chemical and engineering fundamentals of methanol reformers and fuel cells, as well as the factors governing product cost, marketing and business managers will be better able to interface with customers and arrive at a pragmatic product description.

As much as possible, accurate economic information related to the cost of reformers, hydrogen purifiers, and components is included. To the best of my knowledge, this represents the first published discussion of real and actual costs associated with methanol reformers and hydrogen purification. However, the reader should be reminded that economic information may change quickly and the data included herein are only warranted to be accurate as of April 2010.

Despite the slow pace of commercialization, fuel cell companies still enjoy a surprising level of public support and interest. Perhaps one reason is that societies around the world enjoy the use of electrical power in ways never imagined one and two generations ago. One could safely argue that the most developed countries are addicted to electricity, while the least developed countries need more access to electricity to improve the quality of life as well as increase life expectancy. Beginning with the Industrial Revolution we have witnessed an evolution in engine technology and fuels that engines run on. For sure, engines are great for converting chemical energy to mechanical energy, such as is needed for motive applications. But mechanical energy offers less utility in the Digital Age, where we enjoy and rely on a vast assortment of solid-state devices that require DC electrical power.

The generation of electricity by the conventional two-step processes of first converting chemical energy to mechanical energy and then converting mechanical energy to electricity is becoming obsolete. Over the next few decades we should expect and demand that a greater portion of the electrical power we use daily be produced by more efficient and/or versatile one-step processes such as fuel cells and solar generation. Fuel cell technology provides a one-step process for the conversion of chemical energy directly to DC electrical energy. In this sense, fuel cells may be thought of as batteries that never run down or need charging provided fuel is maintained in the fuel tank.

At the time this text is being written, fuel cell systems operating on reformed methanol still have not been broadly commercialized, but commercial traction is gaining. Although most of the technological barriers have been removed, successful products must also be competitively priced. Although engineers and scientists may find new technology exciting and worthy in its own right, customers will always, to a large degree, make their purchasing decisions on the basis of price. Succeeding at developing new technology is only a part of the battle; to win, products must also be affordable.

April 2010 *Dave Edlund*

Chapter 1

HYDROGEN FUEL CELL TECHNOLOGY

Sir William Grove is usually credited with inventing the hydrogen fuel cell in 1845 based on earlier observations of Friedrich Schoenbein (first published in 1839).[1] This pioneering work by Schoenbein and Grove utilized hydrogen as the fuel and oxygen from air as the oxidizer, and the fuel cells operated close to room temperature. We now know that many different fuels (e.g., hydrogen, hydrazine, methanol, formic acid, and ammonia) and oxidizers (e.g., air, pure oxygen, hydrogen peroxide) may be employed in fuel cells,[2] and fuel cells may operate over a broad range of temperatures up to about 1000°C.[3] However, this work will focus attention on hydrogen–air fuel cells that operate at temperatures slightly above ambient: so-called proton exchange membrane (PEM) fuel cells.

1.1 Common Classifications

Several different types of fuel cells have been the subject of extensive research and development in recent decades. Initially driven by the space race and the need for electrical power generation on space craft, the development of fuel cell technology and systems has more recently been driven by commercial applications.

There are six principle classifications[4] of fuel cells summarized in Table 1.1. These classifications are based primarily on the general

---

Methanol Fuel Cell Systems: Advancing Towards Commercialization
By Dave Edlund

www.panstanford.com

Table 1.1 Classification of common fuel cells types

	Electrolyte	Approx. Operating Temperature	Fuel
PEMFC	Nafion	−20°C to 80°C	High-purity H_2
DMFC	Nafion	−20°C to 60°C	High-purity methanol
AFC	Potassium hydroxide solution	0°C to 80°C	High-purity H_2
PAFC	Phosphoric acid	180°C to 220°C	H_2, CO
MCFC	Molten metal carbonates	400°C to 550°C	H_2, CO
SOFC	Ceramic oxide-ion conductors	600°C to 1000°C	H_2, CO, CH_4

Abbreviations: PEMFC = proton-exchange-membrane fuel cell; DMFC = direct-methanol fuel cell; AFC = alkaline fuel cell; PAFC = phosphoric-acid fuel cell; MCFC = molten-carbonate fuel cell; SOFC = solid-oxide fuel cell

composition of the electrolyte layer that separates anode from cathode. Fuel cell types may also be classified based on operating temperature and type of fuel that is required for satisfactory operation. As can be seen in Table 1.1, inorganic electrolytes provide for higher operating temperature; and higher operating temperature allows for fuel other than hydrogen to be consumed at the anode.

Most commercialization work has focused on PEMFC and SOFC technologies. In particular, the auto industry favors PEMFC over other fuel cell classifications, and stationary power generation seems to favor SOFC, although both FuelCell Energy (USA) and MTU (Germany) have made good progress with large-scale MCFC stationary combined heat-and-power (CHP) systems. It remains to be seen if MCFC technology will remain commercially competitive as the requirement to circulate the molten electrolyte salt imposes some severe material and engineering challenges. The PAFC realized early commercial success largely through sales of the PC-25 200 kW combined heat-and-power plant sold by United Technologies Fuel Cells.

Historically, AFC technology has been extremely important in space exploration. Indeed, AFC systems (a total of three power plants) are still in use on every space shuttle. AFC technology relies on an alkali-metal hydroxide in aqueous solution (commonly KOH), which is continuously circulated through nickel gauze as the electrolyte layer. This circulating liquid electrolyte also serves as the coolant. AFC technology has failed to generate substantial commercialization interest mostly due to the incompatibility of this type of fuel cell with CO_2 (note that on space craft, liquid O_2 and H_2 are available, so there is no issue with CO_2 contamination). Even low concentrations of CO_2 as is found in ambient air (approximately 350 ppm CO_2 on average) can lead to precipitation of alkali metal carbonates in the electrolyte with resultant plugging of tiny flow passages that constitute the nickel gauze electrolyte layer. Even worse, it is extremely difficult to imagine a hydrocarbon or alcohol reformer coupled to an AFC in any practical, cost-effective fashion because of the high concentrations of by-product CO_2 formed when carbon-containing fuels are reformed.

Closely related to PEMFC technology is the DMFC type of fuel cell. The principle difference is that PEMFCs utilize H_2 as fuel, whereas DMFCs use a dilute stream of methanol in water. Although methanol does catalytically react in the presence of platinum group metals at the anode, the rate of reaction is slow relative to the reaction rate for hydrogen. Thus, a substantially higher degree of anode catalyst loading is done with DMFC cells to provide more surface area for methanol reaction. In addition, DMFC technology has not reached a competitive state with PEMFC in terms of power density (W/cm^2). This places DMFC technology at a severe cost disadvantage when compared to PEMFC. Nevertheless, DMFC *systems* appear to hold an advantage over PEMFC *systems* for low-power consumer electronics and remote sensors where small system packaging combined with the convenience of a liquid fuel outweigh the loss in stack power density.

1.2 PEMFC Construction and Basic Principles of Operation

There are many similarities in construction and operation between a PEMFC stack and a battery. There are also several differences (see Table 1.2). It is often helpful and instructive to think of a fuel cell in

Table 1.2 Comparison and contrast of batteries and PEMFC stacks

	PEMFC Stack vs. Battery
Similarities	1. Both produce DC electrical power 2. Voltage "sags" at high current draw 3. Higher voltage is achieved by arranging electrochemical cells in series 4. Both use an appropriate electrolyte layer to separate the anode and cathode
Differences	1. Fuel and oxidant are internal to batteries; external to PEMFC stacks 2. Batteries self-discharge upon storage; PEMFC stacks do not self discharge 3. PEMFC stacks require auxiliary components to supply a flow of air and H_2 to the stack; batteries are completely self-contained

terms of a battery that simply has its fuel and oxidant stored externally (fuel and oxidant both are always stored internally to batteries). This is especially true when considering the direct-current electrical output from a fuel cell (either individual cell or stack of cells*) as batteries also produce direct current electrical power. For instance, a collection of nominally identical batteries (all at the same state of charge and based on the same electrochemistry) may be electrically wired in series to produce higher voltage while maintaining the current output of a single cell, or in parallel to produce higher current while maintaining the voltage of a single cell. The same is true for a collection of nominally identical individual fuel cells.

Most often, individual fuel cells are arranged in electrical series (typically by building the fuel cell stack using bipolar current collection plates) so that voltage is maximized and current is held to minimum values. The practicality of this approach is dictated by the expense associated with large cables and conductors necessary for high-current, low-voltage devices. An exemplary PEM fuel cell stack

* A comment on fuel cell notation is warranted. A single cell will be identified as such, and a collection of multiple fuel cells will be termed a stack. Unfortunately, the term "fuel cell" as commonly used is ambiguous and may be used to refer to either a single cell or a stack.

is shown in Fig. 1.1. This stack consists of 36 individual electrochemical cells in electrical series.

The analogy with batteries goes further. If alternating current is required (often the case when household and industrial appliances are to be powered), then a commercial inverter is simply connected to the direct-current output of the fuel cell. However, since commercial inverters have been designed and manufactured assuming common battery bus voltages (e.g., 12 VDC, 24 VDC, and 48 VDC), this implies that fuel cell systems that are intended to serve in markets where the user may want to power a commercial inverter should be designed to emulate common battery bus voltages, and in fact they do. Commercial power electronics (inverters and DC/DC converters) are completely compatible with fuel cell systems producing electrical power in the voltage range and power range specified for the power electronics.

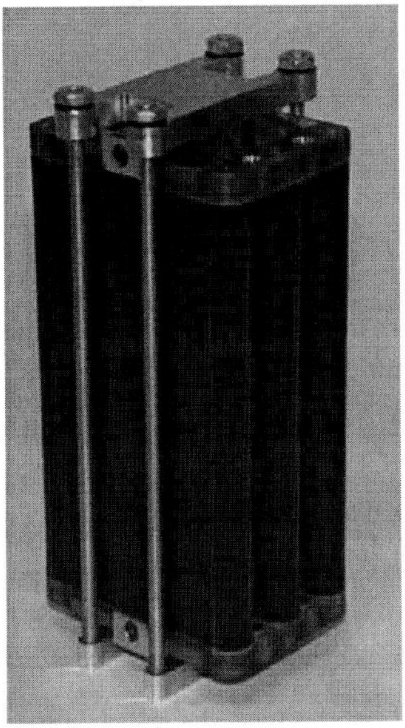

Figure 1.1 A representative PEM fuel cell stack rated to deliver a maximum of 600 W and nominal steady-state output of 310 W at 24 VDC. Fluid connections are made to the amber-colored plastic end plates. Photo courtesy of Protonex Technology Corp. See also Color Insert.

The electrochemical cell common to the PEMFC is very simple in its gross features, consisting of a water-swollen cation exchange membrane electrolyte layer that separates a catalyzed anode from a catalyzed cathode (Fig. 1.2). Finely divided platinum on carbon is the typical catalyst for both anode and cathode, with loadings of about 0.1 to 0.2 mg Pt/cm^2 on the anode and 0.2 to 0.3 mg Pt/cm^2 on the cathode. The higher Pt loading on the cathode is because the cathode reactions (O_2 reduction) are slow relative to the anode reactions (H_2 oxidation). This combination of electrochemical catalyst deposited onto the electrolyte layer is called a membrane electrode assembly (MEA).

However, there are many subtle features that are essential to achieving high performance from a PEMFC. As one might imagine, it is essential that electrons are conducted with minimal electrical resistance from the anode, through the gas diffusion layer (GDL) and into the current collector plate. This, in turn, implies that surfaces are free of oil and other residues, and oxidation, which often leads

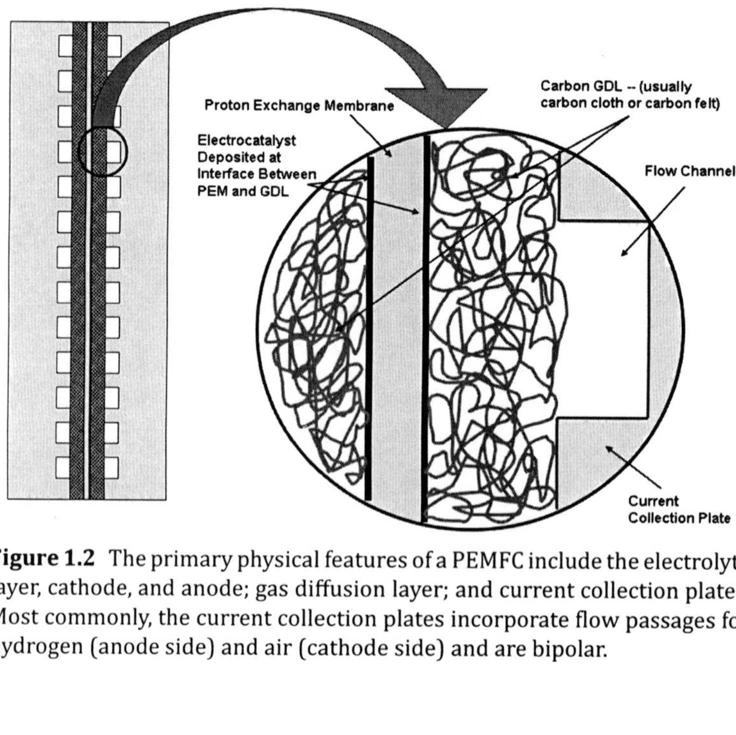

Figure 1.2 The primary physical features of a PEMFC include the electrolyte layer, cathode, and anode; gas diffusion layer; and current collection plates. Most commonly, the current collection plates incorporate flow passages for hydrogen (anode side) and air (cathode side) and are bipolar.

to poor electrical contact. Furthermore, it is also important that the electrocatalyst particles (e.g., Pt on carbon) contact both the electrolyte layer and the GDL.

The ideal electrochemical reactions are

$$\text{Anode: } H_2 \rightarrow 2H^+ + 2e^- \tag{1.1}$$

$$\text{Cathode: } O_2 + 4H^+ + 4e^- \rightarrow 2H_2O \tag{1.2}$$

$$\text{Net: } 2H_2 + O_2 \rightarrow 2H_2O \tag{1.3}$$

The net reaction reflects the fact that the PEMFC derives electrical energy derived from the combustion of H_2. The elegance of the fuel cell is that it decoupled the half reactions using an electrolyte membrane so that a major portion of the combustion energy is derived in the form of electrical power. Some heat is also unavoidably produced as a by-product. The electrical efficiency of the PEMFC is expressed as

$$\eta_{FC} = \frac{V_{cell}}{1.23\,V} \tag{1.4}$$

where η_{FC} is the electrical energy efficiency of the cell, V_{cell} is the measured cell electrical voltage, and $1.23\,V$ is the maximum theoretical cell voltage[4] (under normal operating conditions). It is very common to achieve acceptable power density (W/cm^2) at cell potentials in the range of 0.65 to 0.75 V yielding stack electrical energy efficiencies in the range of 52.8% to 61.0%.[†]

The electrolyte layer is most commonly Nafion®, a sulfonated perfluoropolymer originally invented by DuPont but now sold by other companies as well. The sulfonated groups are hydrophilic, whereas the perfluoropolymer backbone and side chains are hydrophobic. The result is that the polymer adopts a micelle structure incorporating the sulfonated groups in water-swollen regions. It is important that the electrolyte be a good conductor of protons but an electrical insulator. The water-swollen micelles incorporating the negatively charged sulfonated groups provide this ionically conductive pathway,

[†]The difference between 100% and the percent electrical energy efficiency is the fraction of energy released from the PEMFC stack as heat.

but only if water is present. Thus, maintaining an adequate degree of water in the electrolyte membrane is important to achieving high performance from a PEMFC.

Universally, a porous carbon layer is placed on the side of the anode and cathode opposite the electrolyte layer. This layer, called the gas diffusion layer or GDL, tends to restrict the diffusion of water vapor away from the MEA to reduce water loss from the MEA. A second purpose of the GDL is to provide a continuous, electrically conducting medium in contact with both the electrodes (anode and cathode) and the bipolar plates (current collection plates). The GDL may be a non-woven carbon paper or felt, or a woven carbon cloth. The thickness, porosity, and surface treatment (if any) are usually proprietary to the manufacturer and are selected to provide varying degrees of water vapor transport resistance. Also, the degree of compression applied to the GDL during assembly of a cell or stack of cells is critical. Generally, more compression is expected to improve electrical contact and to reduce resistance losses, but too much compression eliminates much of the void space in the GDL and gas transport may be impeded.

This summarizes the general construction of a PEMFC and basic principles of operation. There are two basic types of PEMFCs for operation with hydrogen (or hydrogen-rich) fuel and air. They are classified as low-temperature PEMFC (LT PEMFC) and high-temperature PEMFC (HT PEMFC) since they differ in acceptable range of operating temperature.

1.3 Low-Temperature PEMFCs

Operating typically within the range of about 50°C to 80°C, LT PEMFC is by far the most studied and developed of the commercially favored fuel cells. It utilizes a water-swollen MEA, typically Nafion or equivalent. Examples of LT PEMFC stacks include those under development by Ballard Power Systems, United Technologies Fuel Cells, Protonex Technology Corporation, ReliOn, Inc, Nuvera Fuel Cells, and Chung-Hsin Power Systems, to name a few. Worldwide, there are many companies, large and small, engaged in developing and commercializing LT PEMFC technology, including many major

Table 1.3 Comparison of LT PEMFC to HT PEMFC

	Advantages	Disadvantages
LT PEMFC	1. Rapid cold start 2. High power density 3. Good durability	1. Requires high-purity H_2 2. Water balance may be difficult to maintain
HT PEMFC	1. Accepts low-purity H_2 2. Insensitive to water balance	3. May absorb water at $<100°C$ 4. Must be heated to $>140°C$ at start-up

manufacturers of consumer electronics and most automobile manufacturers.

All LT PEMFC stacks are characterized by a fundamental requirement for high-purity H_2 fuel. This requirement is a direct result of the low operating temperature of the stack. At relatively low temperatures, the anode electrocatalyst (commonly Pt on carbon) is subject to poisoning—reversible and irreversible—due to contaminants including carbon monoxide (CO), unsaturated hydrocarbons, and organosulfur compounds.

As is often the case, there are advantages and disadvantages to both LT PEMFC and HT PEMFC, so it is important to understand how they compare (a summary is provided in Table 1.3). The reader is referred to other publications for a comparison of PEMFCs (both LT and HT) to other types of fuel cells.[3,4]

1.3.1 *The Importance of Operating Temperature*

The operating temperature of a PEMFC (either LT or HT) is an important factor, but one must be careful to be clear as to where the temperature is being measured (reported) and under what conditions. Clearly, since proton conduction through the MEA of all LT PEMFC must occur for useful current to be generated, the practical upper operating temperature is limited by the concentration of water that remains in the MEA. The impact of too little water in the electrolyte layer is severe, potentially causing irreversible failure of the electrolyte (see Fig. 1.3). Too much water may result in localized blockage of the cathode (a phenomenon called flooding) but usually does not lead to irreversible cell damage.

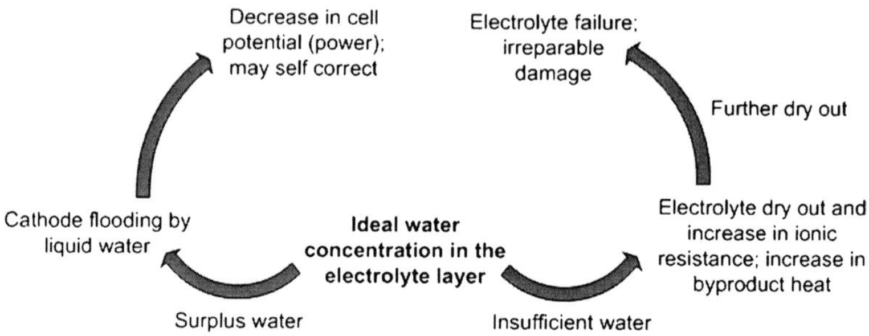

Figure 1.3 A schematic representation of the likely effects of too much or too little water in the PEMFC electrolyte layer.

Given the importance of dissolved water in the electrolyte layer to high-performance operation of the fuel cell, the temperature of greatest importance is the temperature of the electrolyte layer. However, in practice this temperature is very difficult to measure, and it varies over the surface of the MEA (power density also varies over the surface of the MEA). So for practical reasons, the operating temperature of a PEMFC is usually measured by the temperature of the cooling fluid (liquid or air) where it exits the stack. The stack temperature, then, is easily controlled off this measured fluid temperature knowing that the hot fluid exiting the stack is actually cooler than the hottest portions of the stack. This works fine, but it does not tell us the temperature of the electrolyte layer.

To determine the acceptable operational temperature range for a given fuel cell stack, one needs to operate the stack over a range of temperatures and determine at what elevated temperature performance (power) begins to fall off. Some loss in output power with increasing operating temperature is acceptable and may be expressed as a temperature de-rating. However, as the operating temperature is further increased, a point will be reached where output power decreases substantially and continued operation will cause failure within hours. Clearly, this is not an acceptable operating temperature for the fuel cell stack. The challenge is that stack performance as a function of temperature lies on a continuous curve, meaning that the maximum stated operating temperature for a LT PEMFC is a balance of risk aversion and the desire to have broad operational limits.

1.3.2 *Operational Durability*

Most commercial applications of PEMFC systems will require an operational lifetime of at least 500 hours, and for many, >5,000 hours will be required. A convenient reference point is a hypothetical automobile fuel cell propulsion system. Assuming the automobile is operated at an average speed of 50 miles per hour and is driven 150,000 miles, it will have been operated for somewhat more than 3,000 hours. Most consumers would probably find such a short lifetime objectionable; however, many original car owners will have sold the automobile prior to reaching 150,000 miles.

The most demanding applications (with respect to operational lifetime) are combined heat-and-power stationary applications, where the fuel cell system is intended to provided primary electric power and byproduct heat to a residence or light commercial installation. In this class of applications the operational lifetime needs to exceed 15,000 hours and most likely needs to be as much as 40,000 hours (there are 8,766 hours in a year; 40,000 hours of operation is only 4.56 years).

LT PEMFC uses a ruggedized and durable MEA provided water management is adequate. W.L. Gore, DuPont and 3M have all developed mechanically reinforced sulfonated perfluoropolymer membrane electrolyte thin films for fuel cell applications. Combined with chemically stabilized formulas that are more resistance to oxidative degradation, MEAs made using these new state-of-the-art electrolyte layers have demonstrated operational lifetimes in excess of 15,000 hours under controlled testing. Equally important, these electrolyte films are tough and readily handled without tearing or breakage.

1.4 High-Temperature PEMFCs

PEM-type fuel cells that do not use a water-swollen polymer electrolyte membrane are grouped into a classification known as HT PEMFC. A proton-conducting compound other than water is necessary. The prototypical example is the PEMEAS Celtec-P® MEA (now sold by BASF), which incorporates a polybenzimidazole (PBI) membrane and is swollen with phosphoric acid. This was commercially introduced in 2004 and is targeted at operation within the range of 120°C to 180°C.

Early work with poly(2,2′-m-phenylene-5,5′-bibenzimidazole) doped with varying amounts of phosphoric acid demonstrated a clear relationship between the concentration of acid and the ionic conductivity of the resulting membrane: higher acid concentration leads to higher ionic conductivity. However, simple polymerization of m-PBI followed by acid doping results in poor mechanical strength at necessarily high acid concentrations. Furthermore, the acid is susceptible to leaching out of the membrane if exposed to liquid water.

To overcome this problem, the Celtec-P membrane is made by an innovative process in which the m-PBI membrane is made using a sol–gel process conducted in polyphosphoric acid. *In situ* hydrolysis of the polyphosphoric acid following membrane casting yields a high concentration of phosphoric acid within the membrane; up to 30 acid molecules per repeat unit is reported while still maintaining acceptable mechanical properties.[5] Exposure to liquid water and the resulting leaching of phosphoric acid from the membrane are still a concern.

The major advantage of HT PEMFC designs is that it can tolerate a substantial concentration of CO in H_2 fed to the anode. The anode and cathode catalysts are Pt on carbon, same as for the LT PEMFC. The specific degree of CO tolerance is a strong function of the temperature at the anode. Acceptable electrochemical performance is obtained with up to 1% CO at 140°C, but this increases to 3% CO at 180°C.[6]

Another advantage of HT PEMFCs is that there is no need to humidify the cathode air supply. Since the electrolyte layer does not require water to maintain good proton conductivity, the humidity of the cathode and anode streams is irrelevant. However, as mentioned above, it is important that contact between liquid water and the MEA be avoided so that phosphoric acid is not leached from the MEA.

At present there are two major drawbacks of HT PEMFC in general, and phosphoric acid–doped PBI in particular. The most obvious is the high operating temperature. This requires that the fuel cell stack be heated from ambient temperature (this would normally be in the range of −20°C to +45°C) to the minimum operating temperature: 120°C for Celtec-P MEAs. The time and energy required to heat a fuel cell stack is dependent, of course, on the size (mass) of the stack, among other factors. Suffice it to say that the time delay associated with heating the stack is a significant disadvantage for all applications where instant (or at least rapid) start-up is a requirement. But for applications where a time delay on start-up is acceptable, such as stationary CHP, HT PEMFCs may be a good choice.

High-temperature fuel cells using phosphoric acid–doped PBI MEAs are basically phosphoric acid fuel cells. As such, they are subject to similar electrode chemical reactions. In particular, the cathode activation polarization is significant with phosphoric acid electrolyte. The result is that, in practice, the practical limit on cell performance is about 200 mA/cm^2 at 0.7 V/cell, and about 500 mA/cm^2 at 0.6 V/cell using H_2 and air. If reformate containing 2% CO is used as the fuel gas rather than H_2 (a more reasonable assumption) the performance decreases to about 100 mA/cm^2 at 0.7 V/cell, and about 300 mA/cm^2 at 0.6 V/cell.[6] In comparison, a LT PEMFC is expected to produce about 670 mA/cm^2 at 0.7 V/cell—about 6 times higher power output than is obtained with the Celtec-P MEA.

1.4.1 *Future Potential of HT PEMFCs*

As attractive as the advantages of HT PEMFCs are, the future potential of this class of fuel cells remains uncertain. At the present, there is only one commercial source of high-temperature MEAs: BASF (Celtec-P), although other commercial sources are being developed. Consequently, with little competition there is uncertainty as to when and what factors will result in cost reduction of the expensive MEA.

Two potential major customers for Celtec-P—Plug Power and Motorola—have seemingly lost interest. Motorola has completely abandoned its micro-fuel cell program that was aimed at coupling a miniature methanol reformer directly with a HT PEMFC. Plug Power and partner Vaillant GmbH have failed to deliver a commercial residential CHP system using the HT PEMFC stack.

However, for CHP applications ClearEdge Power continues to develop a multi-kilowatt HT PEMFC system using the Celtec-P MEA, and 5 kW commercial systems are being deployed in California. For micro-portable power, UltraCell continues to develop a HT PEMFC product rated at 20 W (model XX25). Testing of prototypes provided to the United States Army has generally provided good results.

For now it appears that the major effort on further developing high-temperature MEAs is being done in Europe through the CARISMA program.[7]

1.5 System Architecture and Balance-of-Plant

So far we have mostly limited the discussion to the fuel cell stack and internal components. However, the PEMFC stack cannot be operated without a supporting balance-of-plant (BoP) that provides certain functionality as listed in Table 1.4.

The necessary functional requirements given in Table 1.4 are fairly easily achieved. For instance, a cylinder of compressed H_2 and a needle valve may be considered adequate for supplying a regulated flow of fuel to the anode. Likewise, a blower or air pump may suffice for supplying clean cathode air, especially if combined with a mass-flow sensor, feedback control loop, and means for varying the output of the pump. In fact, it should be pointed out that if the stack is only required to operate at very low power density, then natural convection and diffusion may be adequate for supplying the cathode with oxygen. In this special case, no mechanical air pump (or fan or blower) is required.

Table 1.4 Necessary and optional functional requirements of the fuel cell BoP

Necessary	Optional
1. Regulated fuel flow to the anode side of the stack 2. Regulated air flow to the cathode side of the stack 3. Adequate filtration of the cathode air 4. Adequate cooling of the stack (liquid or air)	1. Humidification of the cathode air 2. Humidification (recirculation) of the anode gas 3. Fuel cell stack output current control 4. Over temperature protection 5. H_2 leak detection and safe shutdown 6. Microprocessor-based controller

The value in providing the optional functionality (Table 1.4) may not be obvious. The need to humidify the cathode air is dictated by the performance of the MEA plus the intended operating conditions. As an extreme example, almost any commercial MEA will operate satisfactorily if the current density is low; e.g., less than 100 mA/cm^2. However, seldom is this desirable or economical, so humidification is usually thought of as required BoP.

Humidification of the anode gas, especially by recirculating the anode gas, can be a valuable supplement to cathode air humidification. This is seldom needed in closed-cathode PEMFC stacks, but open-cathode stacks can benefit greatly from recirculating the anode purge gas. It is common to bleed 10% to 20% (maximum) of the anode gas from the recycle loop to avoid accumulation of water and N_2.

If a microprocessor-based fuel cell controller is used, then the other optional functional features may be provided. The potential to limit, or clamp, the output current from the stack is a useful method of protecting the stack from excessive current draw by the load and the resulting collapse of cell voltages. Over-temperature protection is another optional measure to protect the stack from dry out and irreversible damage. Hydrogen leak detection tied to a control loop enabling safe shutdown is purely to address safety concerns.

Most PEMFC stacks are liquid cooled, but some are air cooled. The primary functional differentiation is that a liquid-cooled stack will operate at higher power density than will an air-cooled stack. However, this increase in power density comes at the expense of a more complex and heavier BoP. At a minimum, the liquid-cooled stack will require a radiator (liquid-to-air heat exchanger), a coolant reservoir, and a coolant circulation pump. Normally a fan would be used to direct ambient air flow through the radiator to effect cooling of the heat transfer fluid, but other means could be employed. For instance, an aircraft could make use of movement through the air to effect flow of air over a radiator.

A typical design for a liquid-cooled PEMFC system is presented in Fig. 1.4. Historically, de-ionized water has been used as the heat transfer fluid; more recently a mixture of water with either propylene glycol or ethylene glycol has been used. Another promising heat transfer fluid is a proprietary nanoparticle-stabilized mixture sold by Dynalene, Inc. (Whitehall, Pennsylvania, USA).

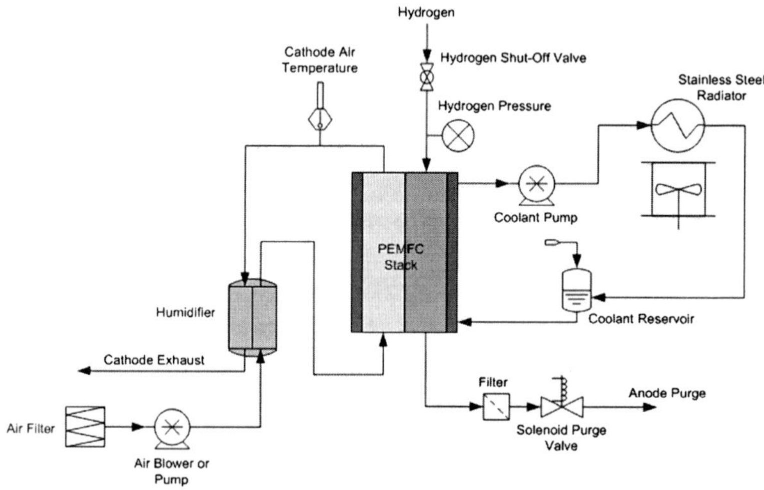

Figure 1.4 Simplified schematic showing balance-of-plant and instrumentation typical for a liquid-cooled PEMFC system.

Regardless of the choice of heat transfer fluid, it is extremely important to use only fluid-wetted materials that are compatible with the heat transfer fluid. This has presented challenges with cooling systems using de-ionized water, and stainless steel or plastic are preferred materials. Brass, copper, and aluminum all corrode quickly in de-ionized water. Even worse, the dissolved metal ions are prone to plating out inside the cooling manifold and coolant passages of the fuel cell stack. When this occurs, the stack will fail prematurely either due to blocked coolant passage or cell-to-cell shorting.

The cooling loop should function to maintain a high, but acceptable, operating temperature of the stack. This can be done by measuring temperature at a choice of locations including the fuel cell stack, the cooling fluid, and the cathode exhaust air. Since air must be flowing through the cathode during operation of the fuel cell stack, this location of temperature measurement is shown in Fig. 1.4.

The cathode air loop shown in Fig. 1.4 reflects a closed loop, meaning the cathode is fed humidified air from a filtered and regulated supply. This is most common with liquid-cooled fuel cells, although it can also be applied to air-cooled fuel cells. Since ambient air may contain a variety of air pollutants that have an adverse effect on stack performance (such as SO_x, NO_x, hydrocarbons, and particulates), filtration is highly recommended. Particulate filtration is well known, and such filters are commercially widely available.

The combination of particulate filtration with chemical filtration of cathode air for a fuel cell has been patented in the USA by Donaldson Company, Inc.[8] Chemical filtration is accomplished by employing a bed of absorbent material such as activated carbon, impregnated carbon, ion-exchange resin or fibers, activated alumina, molecular sieves, silica, and clays.

Regardless of the selection of filter medium, it should be sized for an appropriate operation period under the worst anticipated level of air contamination. Periodic replacement is likely to be necessary, and the design engineer should specify a filter maintenance interval.

Cathode air humidification can be done by water vapor exchange through a suitable permselective membrane as shown in Fig. 1.4, either configured as flat sheets (perhaps pleated to provide a large amount of membrane area in a small package) or hollow fibers. Other humidification methods that may be successfully used include enthalpy wheels and direct injection of liquid water into the cathode air stream.

In contrast to the liquid-cooled fuel cell, the air-cooled fuel cell is supported by a significantly simpler BoP (see Fig. 1.5). The radiator,

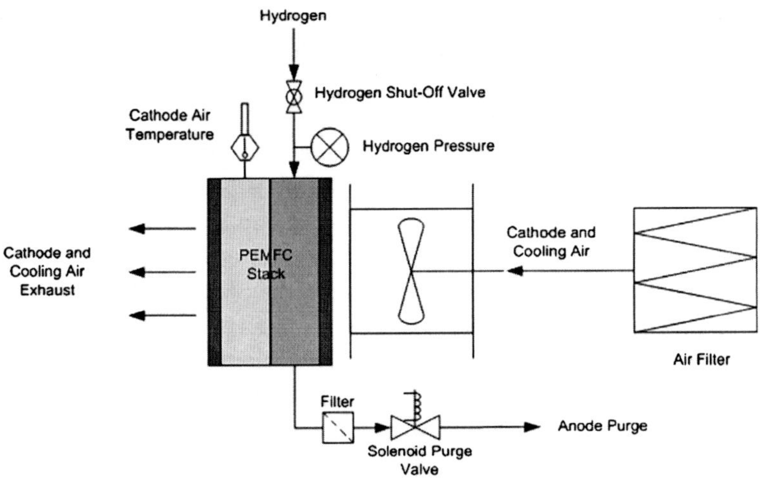

Figure 1.5 Simplified schematic of an air-cooled PEMFC system showing typical balance-of-plant and instrumentation.

coolant circulation pump, and coolant reservoir are eliminated. Since the cathode side of the PEMFC stack is open to ambient air, the functions of cathode air for electrochemical reaction and cooling air are combined—the stack itself is a radiator.

Although the system simplification of the air-cooled stack is attractive, there are some drawbacks to this system approach. As mentioned above an air-cooled stack will not operate at as high a power density as can be routinely obtained with a liquid-cooled stack. This may be an acceptable trade off provided the additional cost associated with the air-cooled stack (required to meet a given power output) is less than the cost savings accrued from a simpler BoP. For instance, a hypothetical liquid-cooled PEMFC stack (sized to produce 500 W net electrical power) may cost $500 and its BoP may cost another $500. If a comparable air-cooled stack plus BoP costs <$1000 to deliver the same 500 W net power, then there is a clear and quantifiable cost advantage to the air-cooled system.

Another disadvantage of the air-cooled PEMFC stack is the open cathode, which is susceptible to contamination from ambient air pollutants. It is also subject to excessive loss of moisture, especially at high altitudes or hot conditions. An air filter can be arranged as shown in Fig. 1.5 to filter all incoming cathode and cooling air, but because the amount of cooling air significantly exceeds the amount of cathode air, the required filter area is greater than it would be for a closed cathode system. A closed cathode system would also address the dry out issues.

Although the maximum degree of BoP simplification is derived from an open-cathode, air-cooled design, it is certainly possible to have a closed-cathode, air-cooled stack and system. In this case, the humidification loop shown in Fig. 1.4 would remain, while the coolant circulation loop shown in Fig. 1.4 would be eliminated in favor of the simple cooling fan approach shown in Fig. 1.5. However, it is difficult to imagine a closed-cathode, air-cooled stack offering any cost advantage over a closed-cathode, liquid-cooled stack for the simple reason that the cost savings attributable to elimination of the coolant circulation loop is more than offset by the additional fuel cell stack cost associated with the low-power density of an air-cooled stack design.

Recirculation of the anode hydrogen stream is an option that may be implemented with any of the fuel cell system designs previously discussed: liquid cooled, air cooled, open cathode, or closed cathode.

Generally there are two reasons that would motivate the system designer to use anode gas recirculation. The first is to increase the flow velocity of hydrogen flowing through the anode flow fields. This may offer improved clearing of water droplets that sometimes form in the anode flow field. The second reason is to provide supplemental humidification.

The anode hydrogen gas stream is normally very dry as it enters the PEMFC. As the gas flows through the anode flow field it picks up moisture and exits with a high concentration of water vapor and, often, liquid water drops. By passing a slip stream of this anode exit gas back to the entrance to the anode flow field, the concentration of water vapor at the entrance is significantly increased and this helps avoid localized excessive drying of the MEA.

For low-power PEMFC stacks, anode gas recirculation is usually not warranted. But if the stack is required to produce kilowatts of electrical power, anode gas recirculation often offers significantly improved longevity and power density.

1.5.1 *Balance-of-Plant Component Selection*

The process of selecting specific components for the BoP necessarily involves many compromises. To be successful, the BoP must be designed, and components selected, on the basis of a detailed understanding of the intended application of the completed fuel cell system. It can be helpful to make a high-level list of key features and attributes of the fuel cell system and then prioritizing them (see examples in Table 1.5). It is not uncommon for certain features/attributes to be mutually exclusive. For example, it is very difficult (perhaps even impossible) to optimize for both minimal cost and maximum durability (expressed as cyclic and operating lifetime). In such case a compromise is called for.

The two hypothetical PEMFC system examples in Table 1.5 illustrate the different degree of importance the system engineer might logically assign to key features and attributes, and how that ranking is influenced by the intended application. In the table, the portable application is assumed to be low power (<300 W) and capable of being moved by one healthy adult. In this application, cost is the most important factor. If low cost is achieved, then customer expectations will likely be relaxed on lifetime. Indeed an operational lifetime for

Table 1.5 Possible ranking of high-level feature/attributes for two hypothetical PEMFC systems targeting different commercial applications

	Priority Ranking (1 = Highest)	
Feature or Attribute	Portable System	Commercial CHP System
Cost	1	3
Fuel Efficiency	3	2
Size and/or Weight	2	4
Long Operating Lifetime	4	1
Long Cyclic Durability	4	2

recreational equipment, a significant market segment for portable fuel cell systems, may never need to exceed 1,000 hours for the vast majority of customers. Furthermore, fuel efficiency is usually not a major driver for recreational and part-time use.

In contrast, a commercial CHP application (assumed >25 kW electric output) will be less sensitive to the purchase price but more sensitive to fuel efficiency since operating costs are reasonably expected to dominate over capital cost. An operational lifetime in excess of 40,000 hours is usually expected (and necessary to justify the investment), but CHP appliances will be infrequently cycled, so the cycle lifetime is expected to be low relative to operational lifetime.

Selecting BoP components and finalizing a system design is not recommended until the above simple exercise has been completed. Also, it is unwise for the engineer to complete this exercise in a vacuum— input from a capable and knowledgeable marketing group is highly recommended to ensure that the best information available about the application is considered.

Once the target application and system design are known, the process of selecting specific BoP components comprises many discrete steps as summarized below:

1. Compile a list of candidate components based on published specifications.
2. Check the candidate components for material compatibility—in particular, be aware of potential off-gassing or leaching of organic compounds from polymers.

3. Refine the list of candidate components and procure samples—ideally a statistically significant sample population is purchased.
4. Conduct durability testing of candidate components under realistic conditions.
5. On the basis of the results from durability testing, down-select specific components.

The first step, compiling a list of candidates, has become relatively easy with the broad adoption of the Internet. Most companies publish their catalogs on the Internet and also provide updates of new product announcements. Clearly, the published specifications for candidate components must satisfy the functional requirements for that component in order to be selected for further evaluation. For example, the published specifications for a candidate cathode air pump should state that the pump meets or exceeds the required air flow rate at the known pressure drop for the cathode air loop. Other specifications such as voltage that a drive motor runs at and seal materials may be viewed as "not fixed," meaning that the manufacturer of that component may be willing to make substitutions even for low-volume orders.

The fluid streams used in a PEMFC system are not especially aggressive chemical compounds or mixtures. Certainly material compatibility concerns with hydrogen and air should be very low. However, since the PEMFC anode and cathode are sensitive to poisoning by volatile organic compounds, care must be exercised to screen materials that might off-gas or, at least in the case of the cathode humidifier, leach foreign materials into water present in the humidifier and cathode air manifold. To this point, one should be suspicious of rubbers and plastics that have a significant concentration of plasticizers (e.g., polyvinyl chloride) or that are made by polymerizing two separate components (such as some epoxies and polyurethanes).[‡]

If the anode stream is to be recirculated, then the anode gas pump must also be evaluated for material compatibility with respect to off-gassing and leaching. In addition, the anode gas pump must also be safe for operation with a flammable gas. This includes a minimum

[‡] These cautions about volatile and/or leachable organics from certain rubbers and plastics applies equally well to tubing that might be used for fluid connections.

amount of hydrogen leakage from the pump and safe, low-voltage or explosion-proof electric motor.

After refining the list of candidate BoP components, samples need to be purchased in order to experimentally validate component durability (operating hours and cyclic). This is a key task since lifetime information published by the manufacturer is unreliable, and often just not available.

1.5.1.1 Statistical analysis and lifetime

Durability testing of components should be done with a statistically significant sample population; 10 to 20 is a reasonable population size, and sample sizes <4 should only be used as a last resort. Endurance testing should be designed and executed to mimic, as much as possible, actual operating conditions including extremes of temperature, pressure, and altitude. Cycling on and off is better than running continuously unless the operating mode of the fuel cell system is continuous.

At the outset, it should be clear exactly what constitutes acceptable operation and failure for each and every component under test. For instance, a cathode air pump should be failed at any time it can no longer provide adequate air flow rate at specified pressure drop required for full, rated fuel cell output. The only exception to this rule is if the system performance is de-rated under certain operating conditions (such as elevated temperature and altitude). The question of de-rating is strictly related to business and marketing—it is not a technical question. But de-rating is a common concept, and it is appropriate to consider how de-rating may be applied to fuel cell systems.

De-rating may also be applied to end-of-life performance. This is a fairly common practice with commercial products. For example, engines are rated for power and torque based on beginning-of-life performance. And rechargeable batteries are rated for run time (stored energy) based on beginning-of-life performance.

Durability or endurance testing of a statistically significant sample population of any BoP component should logically lead directly to a down-select decision. System engineers should be quite comfortable with component-level testing and the statistical analysis of the results. There are commercial software packages[§] that can be used for this purpose. The importance of statistical analysis is to discern between three different categories of failures: (1) infant mortality, (2) end-of-life wear-out, and (3) random failures. A Weibull analysis is effective at this task, but it is also likely that a graphical representation of number of failures vs. time will visually lead one to the same conclusion (provided the sample population is relatively large).

It is important to understand that statistics does not provide detailed information about the mechanisms leading to component failure. Therefore, experimental testing must be employed to determine how long components will operate under the intended environmental conditions.

Infant mortality usually is reflective of poor manufacturing quality: i.e., the component (e.g., humidifier, cooling fan, cathode blower) was assembled poorly or using inferior materials. Infant mortality is seen as a high probability of failure early in the product life and a *decreasing probability of failure as the component ages.* In other words, if the component is defective, it fails soon; otherwise it keeps operating. It is normal to see infant mortality on new components as the manufacturing methods, quality control and assurance methods, and materials are being developed.

End-of-life wear-out is the expected failure method. There will always be a finite service life for any component; every product eventually wears out. End-of-life failure is characterized by an *increasing probability of failure as the component ages.* Usually, mechanical wear (moving parts) leads to an earlier onset of end-of-life wear-out. An important point to remember here is that a longer component life is not always desirable. For instance, if the PEMFC system is only required to operate for 1,000 hours but must be low cost, then it would be incorrect to select a BoP component that has a lifetime of 5,000 hours if a less expensive option is available but has a lifetime of 1,500 hours.

The most troublesome category is the random failure. Fortunately, this is not often encountered. Random failure is characterized by a *lack of correlation between probability of failure and component age.* Indeed, the observation that a component is failing randomly is

[§] Two examples are Relex Reliability Studio and Weibull++ (Reliasoft).

more likely to indicate that the key variables causing failure have not been fully identified, and further understanding is in order.

Figure 1.6 is a plot showing these three different failure modes as discrete line and also the so-called bathtub curve that is a combination of the three failure modes as might be experimentally observed. For mature components, the relative impact of infant mortality and random failure should be exceedingly low, in which case the bathtub curve would be very flat at low and middle lifetime, and very near to zero failure rate.

Failure rates may be expressed as the number of units that fail at a given time (as shown on the y-axis in Fig. 1.6). This is especially convenient for the purpose of analyzing data from a population of identical components that are going through durability testing. But a more common expression of component lifetime is mean time to failure (MTTF). In this case it is assumed that components in the PEMFC system are discarded rather than being repaired. Bear in mind that the same is not true for the PEMFC system (see below). Mean time

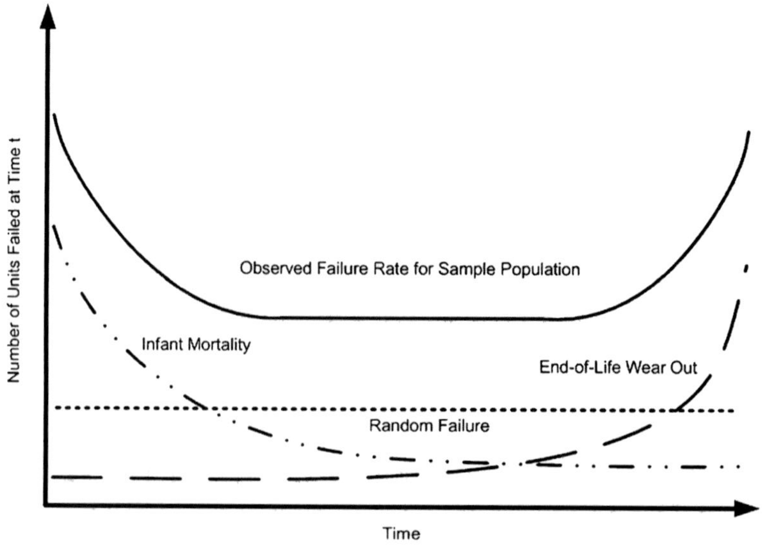

Figure 1.6 Graphical representation of component failure rate as a function of time for a statistically significant sample population tested under nominally identical conditions.

between failures (MTBF) is not appropriate for describing component life expectancy because MTBF assumes the component is repaired and placed back into service.

Remember that the PEMFC system is a collection of parts and components, including the PEMFC stack and many other items. The lifetime of the system is ultimately of principle interest to the customer, the owner of that fuel cell system. It is safe to conclude that the system will cost more than any one of the internal components. So it is reasonable to ask what defines end of life of the fuel cell system. The answer is very much subjective, just as it is if you were to ask what defines the end of life for other products that consumers buy in volume such as appliances, automobiles, and recreational products.

It is likely that fuel cell systems will require repair or servicing to prolong their operational lifetime. The probability that a fuel cell system will need servicing increases with the age of the system. And the probability that the consumer will accept the cost of repair rather than buying a new system increases with the cost of the fuel cell system. This argument suggests that stationary CHP systems are more likely to be serviced than are low-cost portable systems.

The point is that when evaluating the durability of BoP components, one should not lose sight of the fact that most components may be treated as a maintenance or service item. An inexpensive radiator cooling fan is a good example. If such fan costs $20 and the complete PEMFC system costs $1,000, then it would be foolish to declare that the system has reached end of life simply because the fan fails. This discussion is not intended to suggest that the design engineer deliberately down-select BoP components that are likely to fail before the rated life of the system is reached. Doing so would not be a popular move with customers. But, on the other hand, it may be impossible to down-select all BoP components such that they are acceptably priced, meet all functional requirements, and have long proven lifetime.

1.5.1.2 Accelerated lifetime testing

Accelerated lifetime testing is an attractive option to long-term testing of a statistically significant sample population, especially if the goal is to establish a long lifetime. This is especially true in cases where BoP components are targeted for a durability of 10,000 hours to 40,000 hours (one year is 365 days or 8760 hours).

In order to effectively apply accelerated testing, it is necessary to understand the relationship between operational lifetime under normal conditions (i.e., conditions under which the fuel cell system is designed to be used) and under the accelerated test conditions. For electronics, this relationship has been extensively studied and the effect of elevated temperature and humidity on the lifetime of electronic components is understood. However, other components, such as pumps, blowers, fans, humidifiers, etc., may be much more difficult to evaluate under an accelerated test regimen.

Electromechanical components that combine an electric drive motor with a pump, fan, or blower may fail either if the motor fails or if the pump body, fan assembly, or blower head fails. For example, bearings may seize, a pump head may develop leaks or internal check valves could fail to seat, or brushes on an electric motor could wear out. Unless the causes of the failure(s) are known and the mechanism of failure understood, accelerated lifetime testing should not be undertaken.

On the other hand, if the failure mechanism is at least broadly understood, then it might be very productive to design and execute an accelerated test of lifetime. A good example would be wear on a bearing, such as an electric motor bearing and fan bearing. If the operating design point for that bearing requires, for example, a rotational speed of 2,500 rpm, then a reasonable and worthwhile accelerated test would be to spin the bearing at a significantly faster speed. A rotational speed of 10,000 rpm would represent a 4 × acceleration in lifetime testing. This means that each hour of operation under the accelerated test conditions (i.e., 10,000 rpm) equates to 4 hours of operation under the design conditions.

However, it is difficult to maintain all operating conditions at constant values between accelerated testing and normal operation. In the preceding example, it might prove to be impossible to ensure that the bearing is maintained at the same temperature under accelerated testing as normal operation. And if the bearing heats substantially under accelerated testing (for instance, due to friction) then the test results may be influenced. In this example, substantial heating of the bearing would be expected to reduce lifetime, so the actual lifetime predicted by the accelerated testing result may be artificially low.

Another example where accelerated testing may be applied is with diaphragm pumps. This is a common type of pump that may be employed for moving cathode air, recirculating anode hydrogen, and circulating cooling liquid through the fuel cell stack. Diaphragm pumps are known to fail by fatigue of the diaphragm itself. An accelerated test might be to rapidly cycle the diaphragm pump; the total numbers of cycles is counted during the test, and this represents the lifetime. However, localized overheating due to rapid flexing of the diaphragm may skew the results to artificially low values. This is especially a concern for diaphragms made of rubbery materials (polymers).

One of the best applications of accelerated testing is with solenoid valves, for example the anode solenoid purge valve. This valve may be energized every few seconds to minutes in a fuel cell system. If the target lifetime for the fuel cell system is 10,000 hours and the solenoid purge valve is expected to be energized, on average, once every 30 seconds, then over the lifetime of the system it would be energized 1.2 million times. But rather than waiting 10,000 hours to determine if the valve has sufficient durability, one could conduct an accelerated lifetime test of the valve by subjecting a population of 10 to 20 valves to rapid on/off cycling. If one cycle was completed every second, then the accelerated test to 1.2 million cycles would be completed in 334 hours or just under 14 days. And if the accelerated cycling test were conducted on the test population at an elevated temperature that reflects the maximum environmental operating temperature, perhaps 60°C, then the test results are even more convincing.

Notes

1. U. Bossel, *The Birth of the Fuel Cell*, European Fuel Cell Forum (2000)
2. W. Vielstich (translated by D.J. G. Ives), *Fuel Cells: Modern Processes for the Electrochemical Production of Energy*, John Wiley & Sons (1970)
3. A.J. Appleby and F.R. Foulkes, *Fuel Cell Handbook*, Krieger Publishing Company (1993)
4. U.S. Department of Energy National Energy Technology Laboratory, *Fuel Cell Handbook*, 7th edition, prepared by EG&G Technical Services under contract DE-AM26-99FT40575 (2004)

5. B. Du, Q. Guo, Z. Qi, L. Mao, R. Pollard, and J.F. Elter, "Materials for proton exchange membrane fuel cells," in *Materials for the Hydrogen Economy*, edited by R.H. Jones, J.H. Jones, and G.J. Thomas, CRC Press (2007)
6. http:/www.basf-fuelcell.com/en/projects/Celtec-mea.html
7. http://www1.eere.energy.gov/hydrogenandfuelcells/pdfs/htmwg_carisma.pdf
8. A.J. Dallas, M.A. Gogins, K.M. Graham, D. Adamek, and R. Skuster, *Air filter assembly for low temperature catalytic processes*, US Patent #6432177 (13 August 2002)

Chapter 2

METHANOL AS A FUEL

Methanol is the simplest of alcohols: CH_3OH. It is not a hydrocarbon since the hydroxyl group is chemically bonded to the carbon atom. The presence of a carbon–oxygen bond in methanol dramatically alters the properties of methanol relative to its hydrocarbon analog, methane (CH_4). The most obvious difference between the hydrocarbon CH_4 and the alcohol CH_3OH is that methane is a gas under normal conditions whereas methanol is a liquid. What is not so obvious is that methanol is highly polar and reactive relative to methane. Methanol readily adsorbs onto reactive metal surfaces under conditions in which methane is completely inert. Even the C–H bonds are slightly weaker in methanol versus methane; the C–H dissociation energy is about 95 kcal/mole[1] for methanol compared to 105 kcal/mole[2] for methane.

The enhanced chemical reactivity of methanol makes it uniquely attractive as a liquid feedstock for reformers that can be designed to yield hydrogen (via steam reforming) at moderate ranges of temperature; about 250°C to 400°C. In contrast methane steam reforming requires temperatures in the range of 650°C and higher to achieve reasonably fast kinetics.

Historically known as wood alcohol, methanol is formed during the pyrolysis of wood.[3] Modern chemical industries rely on methanol to synthesize formaldehyde, acetic acid, MTBE, and a wide variety of products. Methanol is used to make biodiesel. Fuel cell engineers

Methanol Fuel Cell Systems: Advancing Towards Commercialization
By Dave Edlund

www.panstanford.com

will be familiar with methanol as the anode fuel in direct-methanol fuel cells (DMFC). The PEMFC community has long realized the potential to catalytically reform methanol to yield hydrogen.

The commercial synthesis of methanol is a classic example of chemical engineering. Although most methanol is made from natural gas or methane, other carbon-containing feedstocks may also be used, including coal. In fact, the thermochemical route to methanol is so versatile that methanol has been commercially synthesized from animal waste, garbage, and glycerin (a by-product from the manufacture of biodiesel). In fact, almost any process that can be used to convert carbon-containing material into a mixture of H_2, CO, and/or CO_2 can in principle be extended to the manufacture of methanol.

2.1 Commercial Methanol Synthesis

Worldwide methanol synthesis capacity is approximately 60.8 million metric tons (2008), while actual production is approximately 47.6 million metric tons.[4] Without a doubt it is one of the most important basic chemical commodities. Large-scale commercial synthesis of methanol began in 1923 with the introduction of a high-pressure process using a catalyst based on a combination of zinc oxide and chromium oxide.[5] A more active catalyst comprised of copper and zinc oxide on alumina was introduced in the 1960s and this led to a reduction in operation temperature and pressure. Typical reaction conditions for the synthesis of methanol are about 250°C and 30–100 atm.[5,6]

Most methanol is manufactured from natural gas. Consequently many world-scale plants are located near very large gas fields, especially fields that are remote with respect to gas pipelines. Examples include the Caribbean island nations of Trinidad and Tobago, southern Chile, Western Australia, New Zealand, Indonesia, Malaysia, and the Middle East. At Punta Arenas (southern tip of Chile) Methanex operates four plants with a combined capacity of 3.84 million metric tons annually. And total annual production capacity on Trinidad is 6.58 million metric tons.[7] Methanex Corporation, headquartered in Vancouver, Canada, claims to be the world's largest producer, marketer, and distributor of methanol selling about 7 million metric tons in 2006 (20% of global demand).[8]

Regardless of the basic feedstock used to make methanol, the synthesis loop is essentially the same once synthesis gas is first generated. Synthesis gas is a mixture of H_2 plus CO and/or CO_2.[9] Synthesis gas of an appropriate composition—i.e., sufficient H_2 is present to react with the available CO plus CO_2—is circulated over the catalyst at elevated temperature and pressure. The reaction does not proceed completely to products in a single pass. Methanol is condensed out and the unreacted synthesis gas is circulated back through the loop. Makeup synthesis gas is added to maintain the operating pressure.

The crude methanol from the reactor is distilled to remove water, higher alcohols, ethers, and hydrocarbons that are by-products. A continuous purge from the reactor removes gaseous impurities such as methane, which forms as a by-product.

As mentioned above, the synthesis gas is usually made from natural gas by steam reforming. It may also be made from coke (coal) and wood products. Methane generated by anaerobic decay of garbage, sewage, and animal waste may also be steam-reformed to make synthesis gas.

The primary reactions involved in the manufacture of methanol from synthesis gas are[5]

$$CO + 2\,H_2 \rightarrow CH_3OH \qquad \Delta H°298 = -90.64\ \text{kJ/mol} \qquad \Delta G°298 = -25.34\ \text{kJ/mol} \qquad (2.1)$$

$$CO_2 + 3\,H_2 \rightarrow CH_3OH + H_2O \qquad \Delta H°298 = -49.3\ \text{kJ/mol} \qquad \Delta G°298 = +3.48\ \text{kJ/mol} \qquad (2.2)$$

$$CO + H_2O \leftrightarrow CO_2 + H_2 \qquad \Delta H°298 = -41.17\ \text{kJ/mol} \qquad \Delta G°298 = -28.62\ \text{kJ/mol} \qquad (2.3)$$

Reaction 2.3 is the water-gas shift reaction. It is facile under methanol synthesis conditions and serves to convert CO_2 to CO and vice versa. This is an important step in the overall mechanism since CO is generally considered to be more reactive than is CO_2.

Two issues—concern over greenhouse gas emissions and concern over reliance on imported oil—have begun to influence feedstock selection for methanol manufacture. Perhaps more so than most other commodity chemicals, methanol manufacture is adaptable to a range of feedstocks since synthesis gas, the key ingredient in methanol production, can be made from almost any carbon-containing materials.

In the Netherlands, BioMCN produces methanol from biomass. They offer a "green certificate," which documents the renewable origin of the carbon-containing feedstock. Plant capacity was reported to be 250 million liters annually in 2010.[10] Although biomass is the origin of the feedstock, the specific route that is followed involves glycerin, a by-product from biodiesel production. Other processes involving the direct gasification of biomass, specifically low-grade and waste forest products have been proposed.[11]

The Nobel laureate George Olah has proposed using captured by-product CO_2 as the principal carbon feedstock for methanol manufacture. By adding H_2 extracted from water using electricity produced from nuclear energy, photovoltaic, wind, or even hydroelectric dams, he argues methanol could be manufactured while reducing emissions of CO_2, a greenhouse gas. Olah envisions a future time when methanol is produced from renewable resources and by-product CO_2, the methanol serving as a fuel for fuel cells as well as a combustion fuel either directly or via conversion to higher hydrocarbons (such as via Fischer Tropsch chemistry).[12]

2.2 Physical Properties of Methanol

Methanol is a colorless, flammable liquid. It has a faint alcohol odor. The simplest of alcohols, it is a highly polar organic solvent that is miscible with water in all proportions. Unlike ethanol, methanol does not form an azeotrope with water. The significant properties of methanol are summarized in Table 2.1.

Some important physical properties of methanol/water solutions are summarized below. An equimolar mixture of methanol and water occurs at 64 wt% methanol. This mixture is the ideal methanol-to-water stoichiometry for steam reforming. When mixing methanol with water it is best to do so based on weight percentage rather than volume percentage because the volume of the solution is less than the sum of the separate volumes of water and methanol. A reliable test of methanol/water composition is to measure the solution density.

Table 2.1 Physical properties of pure methanol

Molecular weight	32.04
Specific gravity, 20°C[13]	0.7925 g/cc
Boiling point[13]	64.6°C
Freezing point[13]	−97.8°C
Vapor pressure, 20°C[13]	97 mmHg
Flash point, tag closed cup[13]	12.2°C
Refractive index, 20°C[14]	1.3284 ± 0.0004
Density[14]	6.603 pounds/gallon @ 20°C 6.564 pounds/gallon @ 25°C
Vapor density (air = 1.0)[13]	1.11
Lower explosive limits (vol% in air)[13,14]	6
Upper explosive limit (vol% in air)[13,14]	36.5
Ignition temperature[13]	470°C
Threshold limit value[13,14]	200 ppm (260 mg/m^3)
DOT hazard class[14]	Flammable liquid

For applications in which methanol (with or without water) is to be catalytically reformed to yield a hydrogen-rich reformate stream, it is important to specify the purity of the methanol. Globally, the IMPCA (International Methanol Producers and Consumers Association, Brussels, Belgium) methanol purity specification[15] is a widely recognized reference standard. In the United States, the ASTM D-1152-06 standard[16] may be more widely known. However, the ASTM specification does not limit chloride or sulfur content in the methanol, so the IMPCA specification is preferred.

The IMPCA specification is purposely designed to account for potential impurities that may result from bulk shipment via tanker ships, and yet it draws heavily on testing methods published by ASTM International. In practice, methanol is manufactured to a high degree of purity and only one grade is produced. Methanex Corporation, the largest volume manufacturer of methanol, claims that all of their methanol is produced to meet or exceed IMPCA specifications. Contamination of methanol is most likely to occur when it is transferred from bulk containers to other containers.

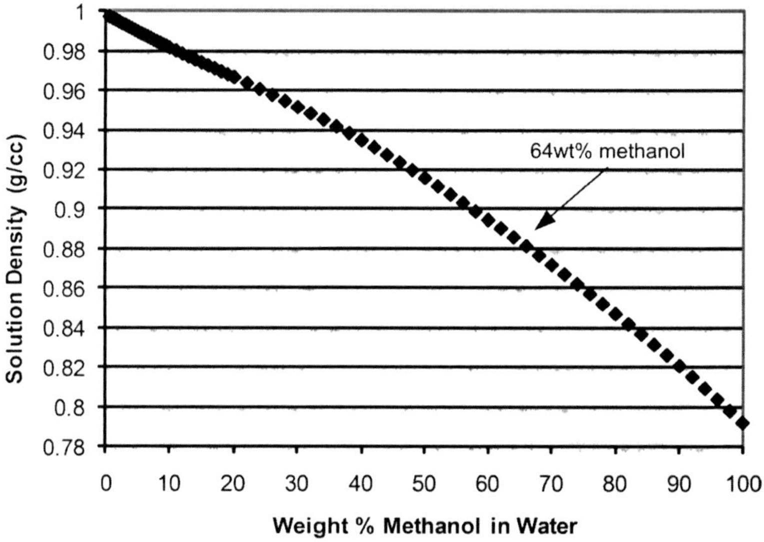

Figure 2.1 Methanol/water solution density as a function of methanol weight percent, measured at 20°C[17]

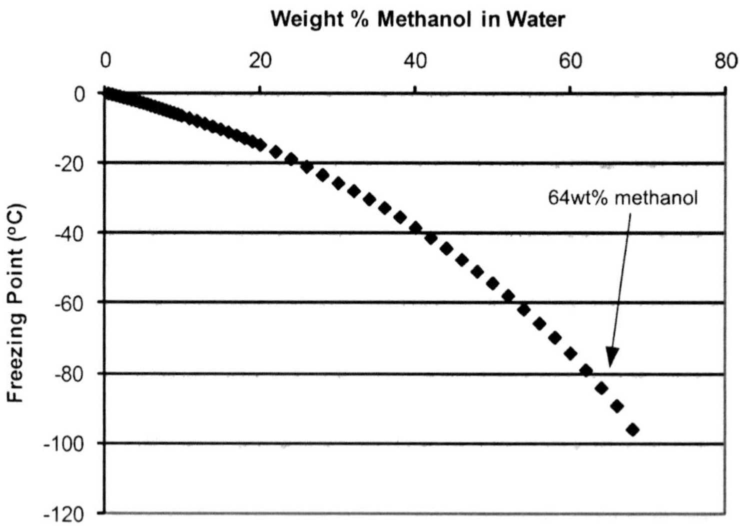

Figure 2.2 Freezing point for methanol/water solutions varying in methanol concentration.[17]

When purchasing methanol in small quantities, chemical suppliers are a convenient source. Larger quantities of methanol may be purchased in drums from chemical and solvent suppliers or fuel distributors since methanol is sold as a racing fuel. Methanol blended with water to a given specification may be purchased from specialty suppliers of solvents and related products. Given the choice, polyethylene and stainless steel are the preferred materials for packaging methanol/water blends.

Table 2.2 ASTM specification for pure methanol

Item	Method	Limit
Apparent specific gravity; 20/20°C		0.7920 to 0.7930
Apparent specific gravity; 25/25°C		0.7883 to 0.7893
Color; Pt-Co, max	D 1209 D 5386	10
Distillation range; max		1.0°C (to include 64.6±0.1)
Non-volatile matter; max	D 1353	5 mg/100 mL
Odor	D 1296	Non-residual
Water, max		0.10 wt%
Acidity (free acid as acetic acid); max		0.003 wt%
Acetone, max		0.003 wt%
Sulfuric acid wash test (carbonizable impurities); color, Pt-Co, max		50
Permanganate time		50 minutes

2.2.1 *Flammability Classifications*

The flammability classification of methanol and methanol/water solutions governs the regulations for packaging and shipping the fluid. This is important to any discussion related to the commercialization of methanol fuel cell systems since fuel must be readily available to commercial customers. And until such time as methanol fuel cell systems are broadly adopted, it stands to reason that the fuel, most

Table 2.3 IMPCA specification for pure methanol

Item	Method	Limit
Appearance	IMPCA 003-98	Clear and free of suspended matter
Purity, min	IMPCA 001-02	99.85 wt%
Acetone, max	IMPCA 001-02	30 mg/kg
Color; Pt-Co, max	ASTM D1209-05	5
Water, max	ASTM D1064-05	0.1
Distillation range @ 760 mmHg, max	ASTM D1078-05	1.0°C (to include 64.6±0.1)
Specific gravity, 20/20°C	ASTM D4052-02	0.791 to 0.793
Permanganate time @ 15°C, min	ASTM D1363-06	60 minutes
Ethanol, max	IMPCA 001-02	50 mg/kg
Chloride as Cl^-, max	IMPCA 002-98	0.5 mg/kg
Sulfur, max	ASTM D3961-98	0.5 mg/kg
Hydrocarbons	ASTM D1722-04	Pass test
Sulfuric acid wash test (carbonizable impurities); color, Pt-Co, max	ASTM E346-03	30
Acidity (free acid as acetic acid), max	ASTM D1613-06	30
Total iron, max	ASTM E394-04	0.1 mg/kg
Non-volatile matter, max	ASTM D1353-03	8 mg/1000 mL

likely a blend of methanol and water, must be shipped directly to the customer or to local retail distribution centers. Certainly the popularity of Internet-based commerce suggests that direct shipments of small quantities of methanol/water fuel, for example one liter to many liters, is a simple and attractive solution to the problem of fuel availability and distribution.

There are two important classification systems to consider when storing and shipping methanol and methanol/water solutions. The United States DOT regulations (the UN DOT regulations are very similar), and the United States National Fire Protection Association (NFPA) regulations. They differ in the classification nomenclature

Table 2.4 Summary of NFPA classifications for flammable liquids[18,19]

Class IA	Flash point <73°F	Boiling point <100°F	
Class IB	Flash point <73°F	Boiling point >100°F	52 wt% to 100 wt% methanol
Class IC	Flash point >73°F	Boiling point >100°F	26 wt% to 52 wt% methanol
Class II	Flash point >100°F	Boiling point >140°F	8 wt% to 26 wt% methanol
Class IIIA	Flash point >140°F	Boiling point <200°F	Up to 8 wt% methanol
Class IIIB	Flash point >200°F		

and rating methodology. The DOT classifies any liquid with a flash point ≤140°F as a flammable liquid. If the liquid has a flash point >140°F then it is classified as a combustible liquid. The NFPA system is more complex and is summarized in Table 2.4.[18]

Pure methanol is classified as a flammable liquid for purposes of shipping. As it is mixed with water, the flash point increases with increasing water concentration (see Fig. 2.3).[20] To be classified as a combustible liquid (as opposed to a flammable liquid) according to

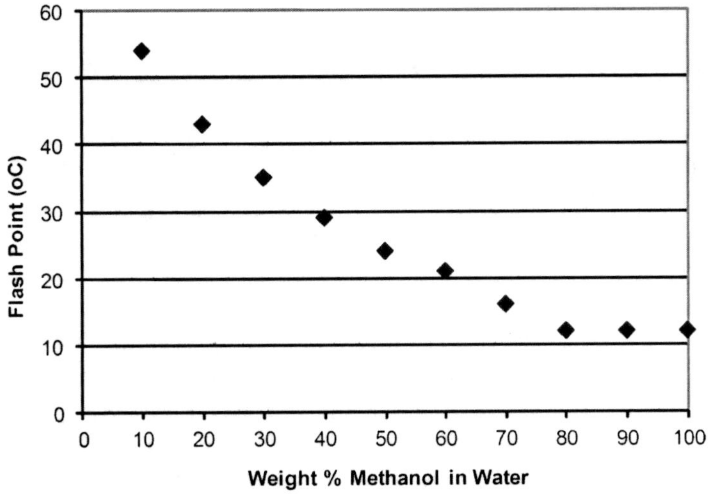

Figure 2.3 Flash point for mixtures of methanol in water.

the DOT system, the methanol/water solution may contain not more than 8 wt% methanol. Approximate methanol concentrations that would fall within each NFPA classification are listed in Table 2.4.

Under the NFPA classification pure methanol is a Class IB flammable liquid. So are most mixtures of methanol and water that are of interest as feedstock for reforming to make hydrogen. Shipping of Class IB flammable liquids in quantities greater than 1 liter, using commercial carriers such as Federal Express and United Parcel Service, requires special packaging in UN approved containers and cardboard cartons. Absorbent material may also be required within the cardboard carton surrounding the container in sufficient quantity to absorb all liquid should a spill occur. Shipping by air is prohibited. However, properly placarded trucks can ship the packaged liquid without absorbent medium within the package.

The German DMFC fuel cell company SFC Smart Fuel Cells has overcome many of the aforementioned shipping limitations for their packaged pure methanol fuel. This was achieved by designing a special fuel container that is tamper resistant and spill proof. The container has been certified to UN and International Air Transport Association (IATA) standards and qualifies for shipment on commercial cargo planes. Two standard sizes are offered: 5 liter and 10 liter.

2.2.1.1 Fuel packaging considerations

In addition to proper packaging of methanol/water fuel to satisfy DOT and/or NFPA flammability classifications, thought should be given to the appropriate size and shape of the fluid container. Materials used for the container and cap must also be considered to ensure compatibility. Generally, polyolefins (polyethylene and polypropylene) are excellent with respect to long-term chemical compatibility with methanol and water, and there are a large variety of affordable containers made from these two polymers.

The filled weight of the container plus fuel should not be excessive since it must be handled by male and female adult customers representing a broad spectrum of physical condition and age. There are no firm guidelines with respect to maximum acceptable weight, but at least one published reference[21] suggests that 35 pounds be the maximum—this is approximately the weight of a 5 gallon container filled with methanol/water solution. However, most would likely consider it easier to pour, without spilling, from a 1 gallon container versus a 5 gallon container.

Small fluid containers avoid weight limitation concerns. However, the size of the container should not be so small as to require a cumbersome number of fuel containers to fill the fuel tank of the fuel cell system. For instance, if the customer is purchasing fuel in 1 liter (or 1 quart) containers but filling a 3 gallon fuel tank, then the customer would be required to open and pour approximately twelve 1 liter containers to fill the tank. This is approaching a degree of awkwardness that many, if not most, customers would prefer to avoid by purchasing fuel in, say, 1 gallon containers.

The fuel container shape should be such as to encourage easy pouring without spilling. A carry handle is a nice feature on gallon and larger-size containers. Square and rectangular bottle shapes pack with less wasted space than do round bottles. This means that the external dimensions of an optimized cardboard carton will be smaller if the bottles, or jugs, are rectangular versus round. The packaging industry has already addressed these concerns for many consumer products, so it is not difficult to find fuel containers that are well suited to the task.

Finally, a child-proof, or tamper-resistant, bottle cap is considered a must given the toxicity of methanol. Even if a taste aversion agent such as Bitrex® is added to the methanol/water fuel mix, a child-proof cap should still be used. And to prevent leaks during shipping, a sealed liner should be applied over the bottle spout (beneath the cap). Aluminum foil liners coated with polyolefin are a good choice. Again, consideration must be given to liner materials that are compatible with methanol and water.

2.2.1.2 Compatible materials

Methanol (pure) and methanol/water solutions are compatible with many polymers, rubbers, some metals, and most glasses and ceramics. There is no reported synergistic effect between methanol and water. Thus, if a material is known to be compatible with pure methanol and with pure water, it will also be compatible with methanol/water solutions at all concentrations.

For seal materials, Viton® is to be avoided because it swells in the presence of methanol. A wetted Viton seal on a moving shaft is likely to seize. Many valves, including solenoid valves, are offered with Viton seals but other materials should always be specified over Viton. Acceptable alternative seal materials are nitrile rubbers, Buna-N, neoprene®, santoprene®, ethylene-propylene, silicones, fluorosilicones, natural rubber, and PTFE.

Hose and tubing materials to avoid include Tygon® tubing, PVCs, and EVA (a blend of polyethylene and vinyl acetate) because the plasticizers present in these rubbers will leach into the methanol or methanol/water solution. Polyurethanes should also be avoided. In the category of metal tubing, stainless steel (series 300 and 400) is preferred, but aluminum is acceptable provided that galvanic corrosion cannot occur. Titanium is compatible albeit expensive. Copper, brass and bronze should be avoided, as well as iron and steel, due to corrosion issues.

Plastics to avoid include polyurethanes, polycarbonates, ABS, epoxies, and nylon. There are many compatible plastics including PEEK, Delrin, EPDM, polyolefins, PPS (Ryton®), PVC, and CPVC. Many useful resources on materials compatibility are available.[22,23]

2.2.1.3 Methanol purity concerns

In the context of using methanol/water solutions as the feedstock for a reformer to produce hydrogen, the methanol and water purity are important considerations. Unlike fuels for combustion, which are relatively insensitive to contamination and impurities, a chemical feedstock should meet or exceed minimum requirements of purity. Common chemical contaminants such as organosulfur compounds and hydrocarbons may adversely affect the activity of the reforming catalyst and/or downstream hydrogen purification unit operations. In the worst case, the catalyst lifetime may be dramatically reduced.

It is not possible to have a detailed and specific conversation about potential contamination species and their effect on catalytic reforming of methanol/water blends without specifying the catalyst and operating conditions. However, some general statements can be made. Reduced sulfur compounds (such as H_2S and organosulfur compounds) and volatile metals (e.g., mercury, cadmium, and zinc) should be considered universal catalyst poisons. Acceptable sulfur levels are typically <1 ppm total sulfur, and 10 ppb is even better. Volatile metals will accumulate on the catalyst over time, so any detectable concentration should be considered less than desirable.

There are many different classes of organosulfur compounds. Thiols and sulfides of the general formula RSH and RSR′ (where R may be the same as R′) are the most reactive class. Heterocyclic organosulfur compounds are less reactive, and the so-called refractory organosulfur compounds, including dibenzothiophene and substituted dibenzothiophenes, are the least reactive class. Generally speaking, the actual degree of poisoning by a given contaminant is expected to be less problematic if space velocities are very high (short residence time).

The impact of temperature on poisoning is difficult to predict. If the contaminant reversibly adsorbs directly onto the catalyst, then increasing the temperature is expected to lessen the effect of poisoning by adsorption. However, if the adsorbed species undergo further reaction on the catalyst surface causing poisoning by formation of secondary species, for instance coking, then increasing temperature may worsen the effects of poisoning.

A generic category of impurity in methanol and water is non-volatile residue. This can be any substance(s) that remains behind when all of the methanol is evaporated. It is measured in units of mg/1,000 mL of methanol and the IMPCA specification places the maximum value at 8 mg/1,000 mL methanol. Non-volatile residue may include silica and salts. Over the operating lifetime of the methanol reformer, the accumulated amount of non-volatile residue that enters the vaporization region of the reformer may be many grams. Reformer vaporizers should be designed with this in mind since an unexpected accumulation of solid residue may impair heat transfer in the vaporizer or, in the worst case, block channels within the reformer.

2.3 Environmental and Safety Issues

The fate of methanol in the environment is governed strongly by two factors: methanol is water soluble and it is naturally occurring. For example, methanol is a by-product of natural fermentation processes and it is also a by-product of the anaerobic decomposition of sewage. Microbial production and decomposition of methanol is ubiquitous.

Normal human exposure to methanol is primarily from consumption of foods and beverages as well as inhalation of atmospheric methanol vapor (either from manmade sources, such as solvents and fuels, or naturally occurring sources). Methanol is found in trace quantities in distilled spirits, beer, and wine, being formed during fermentation of grain and fruits. It is also found in canned fruits and is a metabolic by-product of the artificial sweetener aspartame and the yeast inhibitor dimethyl dicarbonate or DMDC (found in a wide variety of beverages).

A growing market for methanol is to aid in the cultivation of anaerobic bacteria at sewage treatment plants. Addition of methanol to liquid sewage accelerates the growth of bacteria that converts nitrate (itself a by-product of ammonia decomposition) into harmless nitrogen gas. Nitrate pollution of fresh and salt water has been proven to accelerate the growth of algae leading to oxygen depletion. Severe cases of nitrate pollution have had devastating effects on aquatic ecosystems. Over 200 municipal water treatment plants in the United States currently use methanol as an aid for denitrification.[24]

The half-life for methanol spilled on the soil, or released into surface water or ground water, is estimated to be approximately 1 to 7 days. This is a relatively short time frame (similar estimates for benzene suggest a half life of up to several hundred days) due to both the rapid dilution of methanol with local water and rapid bioremediation.[25] Even large spills of methanol into surface water, on the order of hundreds to thousands of gallons, disperse within 1 to 3 days. Many species of methylobacterium and methylomonas isolated from soil will metabolize methanol. Sixty-five different strains of aerobic, gram-negative bacteria have been isolated from sea water, sand, mud, and many aquatic plants that metabolize methanol and are capable of using methanol as their sole source of carbon.[26]

The toxicology of methanol to aquatic life and microbes has been evaluated extensively and many comprehensive and well referenced reports are available.[27,28,29,30] Methanol has low acute toxicity to all species of fish and crustaceans tested; much lower than benzene.[25,26] Furthermore, methanol is not accumulated and concentrated in fish and other aquatic species.[26]

2.3.1 *Safety Hazards Due to Toxicity*

Methanol is toxic to humans if ingested, inhaled, or absorbed through the skin in sufficient quantity. High exposures of methanol can lead to headache, dizziness, nausea, abdominal pain, and impaired vision. An exceptionally comprehensive report on the human toxicological effects of methanol has been published by the United Nations through the World Health Organization.[30]

Excessive exposure to methanol, most likely through either ingestion or inhalation of vapor, has been documented to cause blindness, seizures, coma, and death. Ingestion of 80 mL to 150 mL of methanol is usually fatal to adult humans. Inhalation of methanol vapor in air at a concentration ranging from 4,000 ppm to 13,000 ppm over a twelve-hour period was documented to cause death in one adult worker.[26] Although methanol is absorbed through the skin, typically dermal exposure only leads to mild local drying and irritation. Methanol poisoning is relatively rare in developed countries, but because it initially acts as a central nervous system depressant, much like ethanol, illegal addition of methanol to alcoholic beverages occurs far too often, mostly in developing regions of the world, and most notably Africa and Southeast Asia.

Interestingly, methanol itself is not thought to be toxic in humans and primates. Rather it is the metabolic by-products formaldehyde and formic acid that are cause for concern. The metabolic pathway for methanol in primates and rodents differs. Therefore, methanol toxicity data derived from rodents must be reviewed in detail before extrapolating to humans. In primates, methanol is metabolized in the liver (alcohol dehydrogenase) to formaldehyde which, in turn, is metabolized to formic acid (formaldehyde dehydrogenase). Formate is converted to CO_2 through a mechanism involving the enzyme tetrahydrofolate (THF).[30,31] This last step occurs relatively quickly in rodents, and much more slowly in monkeys and humans. Formic acid also inhibits cellular respiration leading to lactic acidosis.

Interestingly, acute methanol poisoning is treated by one or more of three methodologies. The first, interrupting the conversion of methanol to formaldehyde via alcohol dehydrogenase, is effective if caught very early. Administering ethanol to the patient, which is also metabolized via alcohol dehydrogenase, is one treatment that has proven effective.[32] Alternatively, fomepizole, a potent inhibitor of alcohol dehydrogenase, has also been proposed as early treatment.[33] The second treatment methodology is to counteract the effect of metabolic acidosis, for instance by administering sodium bicarbonate. Finally, administering folic acid, perhaps in conjunction with hemodialysis, can assist with increasing the rate of metabolic conversion of formate to CO_2.

Accumulation of formic acid causes metabolic acidosis and blood pH drops. If untreated severe metabolic acidosis can cause death. Recently, both formaldehyde and formate have been implicated as possible carcinogens in rats fed high levels of the artificial sweetener aspartame daily for the duration of their natural lives (up to 159 weeks).[34] In the cited publication the authors suggest that certain lymphomas/leukemias observed in the female population exposed to daily aspartame doses of 2,500 mg/kg and 5,000 mg/kg may be due to metabolic formation of formaldehyde. However, the authors point out that other observed carcinogenic effects to the renal pelvis and ureter, peripheral nerves, and olfactory epithelium were not observed in long-term assays conducted in their lab using methanol, methyl-*tert*-butyl ether, and formaldehyde. They suggest that multiple metabolites are active and that further research is warranted.[34]

Still, methanol is broadly considered to be non-carcinogenic and non-mutagenic in humans given the lack of reports in the literature that methanol exerts mutagenic or chromosomal effects in humans, and there are no published reports that methanol is a genotoxic to humans. Also, there are no reports of structural alerts, and in vitro studies are negative. At this time the United Stated Environmental Protection Agency has not classified methanol with respect to carcinogenicity.[35]

Many government agencies have published threshold limit values (TLVs) for exposure to methanol in the work place. In the United States, both the Occupational Health and Safety Administration (OSHA) and the American Conference of Governmental Industrial Hygienist (ACGIH) have set the occupational TLV (8 hour daily exposure) at 200 ppm. The World Health Organization published the regulated occupational exposure limits to methanol for most European countries. Although they vary a small amount from country to country, typical time-weighted exposure limits are about 260 ppm.[36]

2.3.1.1 Use of aversion agents

To reduce the risk of accidental poisoning by consumption of toxic household products such as cleaners, automobile antifreeze, and windshield fluid, aversion agents are often added. One such agent is denatonium benzoate, sold under the trade name Bitrex. Manufactured by Macfarlan Smith, now part of Johnson Matthey PLC's Fine Chemicals and Catalysts Division (United Kingdom), Bitrex is claimed to be the most bitter substance yet discovered.[37] As little as 50 ppb in water is said to impart a distinctly bitter taste. The properties of denatonium benzoate are listed in Table 2.5.

Aimed primarily at child safety, Bitrex has been used in commercial products as an aversion agent for over 50 years. It is added to ethylene glycol antifreeze (normally sweet tasting) and windshield cleaner containing methanol at a concentration of 30 ppm. For denaturing ethanol, Bitrex is added to a concentration of 6 ppm to 50 ppm depending on national regulations. Denatonium benzoate is also sold under the trade name ToxiSHIELD™ by Nikita Transphase Adducts Pvt. Ltd. (India). C-Tech Corporation also manufactures and sells denatonium benzoate. There do not appear to be any other commercially-proven aversion agents for use with methanol.

Because of its high molecular weight, denatonium benzoate will not volatilize in the vaporizer preceding a methanol/steam reformer. However, if the vaporizer is heated to 350°C to 400°C, the

Table 2.5 Physical and chemical properties of denatonium benzoate

Property	Value
Molecular formula	$C_{28}H_{34}N_2O_3$
Molecular weight	446.64
Appearance	White, odorless powder or granules
pH (3% solution in water)	6.5 to 7.5
Melting point range	163°C to 170°C
Chloride	0.2% (max)
Loss on drying	1.0% (max)
Solubility	4.5 g/100mL water 69 g/100 mL methanol

ammonium salt will decompose and leave almost no residue. Provided the vaporizer has relatively large passages (not microchannel dimensioned) plugging resulting in high pressure drop will not occur.

However, denatonium benzoate is not compatible with direct methanol fuel cells as it poisons the anode catalyst. Therefore, addition of this proven aversion agent is not possible for methanol fuel to be used in DMFC systems. In that case, special tamper resistant and spill resistant packaging must be employed for safety, usually adding significantly to the cost of the packaged fuel.

Both organic amines and organosulfur compounds have been proposed as potential aversion agents for use with reformed-methanol in fuel cell systems. Neither class of compounds has a commercial track record of performance with methanol; however organosulfur compounds such as alkyl thiols (e.g., ethane thiol, propane thiol, and butane thiol) have been used as an odorant in natural gas and LPG for decades to aid in the detection of minute leaks.

For steam reforming methanol, commercially available catalysts are poisoned by sulfur compounds, but partial oxidation or autothermal reforming are conducted at sufficiently high temperatures that some sulfur tolerance is achieved. Nevertheless, there is no published information on the concentration of alkyl thiol that would be needed to impart an extremely objectionable taste to methanol.

Aliphatic amines are another promising class of potential aversion agents. In low concentrations amines do not inhibit the activity of methanol steam reforming catalysts, and there is no reason to think that POX and ATR catalysts would exhibit sensitivity to poisoning. Biologically amines are one of many degradation products formed when tissue rots. Thus, methyl amine smells strongly of rotten fish. Putrescine (1,4-diaminobutane) and cadaverine (1,5-diaminopentane) are other foul amines of obvious natural origin.

Given the natural tendency to avoid swallowing spoiled food, it is possible that select organic amines would be effective aversion agents. Given the low molecular weight of these compounds, there may be some advantage in vaporizer design and operation if they were used rather then denatonium benzoate. However, alkyl amines, like denatonium benzoate, are not compatible with DMFC anode chemistry or the acidic electrolyte of the MEA.

2.3.2 *Safety Hazards Due to Fire*

When considering the fire hazards of methanol it is helpful to reference gasoline, a flammable fuel that most persons willingly and frequently use on a regular basis. A comparison of important combustion properties of methanol and common gasoline is given in Table 2.6. In general hydrocarbon fuels, gasoline in particular, have much high heat of combustion than does methanol. Also, hydrocarbons burn with an intensely luminous flame and thermal radiation is very high. Methanol, in contrast, burns with a pale blue flame and thermal radiation is very low.

Unlike gasoline and other hydrocarbon-based liquid fuels, methanol fires are effectively extinguished with water because methanol is completely miscible with water. Fire-extinguishing foams may also be used. When fighting methanol fires, the relatively low degree of thermal radiation may allow fire fighters to approach closer to the fire than if it were a combustible hydrocarbon fuel. Thus, one can often approach close enough to use portable extinguishers with good results. Of course, the relatively low thermal radiation generated by a methanol fire also means that there is less threat to surrounding property.

Table 2.6 Comparison of important flammability properties of methanol and gasoline

	Methanol[13]	Gasoline[25]
Composition	CH_3OH	Mix of C4 to C14 hydrocarbons
Specific gravity, 20°C	0.7925	0.72 to 0.78
Boiling point	64.6°C	27°C to 225°C
Heat of combustion	19,930 kJ/kg	43,030 kJ/kg
Flash point, tag closed cup	12.2°C	−42°C
Vapor density (air = 1.0)	1.11	2 to 5
Lower flammability limit	6.0 vol%	1.4 vol%
Upper flammability limit	36.5 vol%	7.6 vol%

The reduced danger of property damage or personal injury from a methanol fire versus a gasoline fire prompted the Indy Racing League (IRL) to require all Indy race cars to burn 100% methanol fuel beginning in 1965. A fiery Crash in the 1964 Indianapolis 500 resulted in two drivers being killed and led to a rule change banning gasoline fuel. In 2006 the IRL moved to 90%methanol/10% ethanol blend, and then 98% ethanol/2% gasoline in 2007.

In moderately cool to cold environments (i.e., <12.2°C), methanol is not flammable unless it is sufficiently heated. As a motor fuel, gasoline is specifically formulated to remain highly flammable even under extremely cold conditions. Methanol vapor is only slightly more dense than air, whereas gasoline vapor is much more dense than air. Thus, gasoline vapor is more likely to accumulate in low areas where ignition sources might be found, rather than to disperse.

The lower flammability limit (LFL) of methanol is 6% in air by volume. This is more than four times higher than the lower flammability limit for gasoline. Since accidental fires usually occur because flammable vapor, as from an undetected leak, increases from a very low level to the minimum concentration at which ignition will occur, the higher the value of the LFL the less likely a fire will result. A fuel with a high LFL value is considered to be safer than a fuel with a low LFL value. Considering all these factors, methanol arguably represents a less severe fire hazard compared to gasoline and other hydrocarbon fuels.

Notes

1. F.R. Cruickshank and S.W. Benson, "Carbon hydrogen bond dissociation energy in methanol," *J. Phys. Chem.* 73, No. 3 (1969) pp. 733–734
2. F.A. Carey and R.J. Sundberg, *Advanced Organic Chemistry*, 5th edition, Springer (2007) pp. 258
3. J. Kemsley, "Methanol's allure," *Chem. Eng. News*, 85, No. 49 (December 3, 2007) pp. 55–59
4. PCI-Ockerbloom & Co., Inc., Global Methanol Supply and Demand, http://www.methanol.org/pdfFrame.cfm?pdf=WrldSD.pdf
5. G.W. Bridger and M.S. Spencer, *Methanol Synthesis, Catalyst Handbook*, 2nd edition, edited by M.V. Twigg, Wolfe Publishing Ltd. (1989) pp. 441–468
6. L.D. Schmidt, *The Engineering of Chemical Reactions*, Oxford University Press (1998) p. 124
7. Jim Jordan & Associates, World methanol Expansion, http://mthanol.org/pdfFrame.cfm?pdf=WorldMethanolPlantsEndof2006.pdf

8. http://www.methanex.com/newsroom/documents/MXGlobalLeader_corporate backgrounder_2007_09_21.pdf#search="producer"
9. G.C. Bon, *Heterogeneous Catalysis: Principles and Applications*, 2nd edition, Clarendon Press, (1987) p. 93
10. http://www.biomcn.eu/news/news/118-biomcn-opens-largest-2nd-generation-biofuel-plant.html
11. K.A. Vogt et al, "Bio-methanol: how energy choices in the Western United States can help mitigate global climate change," *Renewable Energy*, 34 (2009) pp. 233–241
12. G.A. Olah, A. Goeppert, and G.K. Surya Prakash, *Beyond Oil and Gas: The Methanol Economy*, Wiley-VCH (2006)
13. J.R. Couper, O.T. Beasley, and W.R. Penney, *The Chemical Process Industries Infrastructure*, CRC Press (2000) p. 1999
14. Burdick & Jackson Laboratories, Inc., *High Purity Solvent Guide* (1984) p. 74
15. International methanol Producers & Consumers Association, Brussels, Belgium (January 2008), www.impca.be
16. ASTM International, West Conshohocken, Pennsylvania, USA (April 2006)
17. *Handbook of Chemistry and Physics*, 63rd edition, CRC Press (1982) p. D-246
18. R. Burke, *Hazardous Materials Chemistry for Emergency Responders*, 2nd edition, CRC Press (2003) pp. 165–167
19. Methanex Corporation, *Technical Information & Safe Handling Guide for Methanol*, Version 2.0 (October 2002) p. 10
20. http://www.engineeringtoolbox.com/methanol-water-d_987.html
21. W.E. Woodson, B. Tillman, and P. Tillman, *Human Factors Design Handbook: Information and Guidelines for the Design of Systems, Facilities, Equipment, and Products for Human Use* (1992) pp. 623–625
22. http://www.coleparmer.com/techinfo/chemcompresults.asp
23. http://www.efunda.com/DesignStandards/oring/oring_chemical.cfm?SM=none&SC=methanol
24. Methanol Safe Handling Guide, Edition 1.0, prepared by Alliance Consulting International for the Methanol Institute (October 2008), http://www.methanol.org/pdfFrame.cfm?pdf=MethanolSafeHandlingManualOct2008.pdf
25. Evaluation of the Fate and Transport of Methanol in the Environment, prepared by Malcolm Pirnie, Inc., for the Methanol Institute (January 1999), http://www.methanol.org/pdfFrame.cfm?pdf=evaluation.pdf
26. U.S. Environmental Protection Agency, Office of Pollution Prevention and Toxics, Chemical Summary for Methanol (August 1994), Report No. EPA-749-F-94-013a.
27. Methanol in Fuel Cell Vehicles: Human Toxicity and Risk Evaluation, Statoil internal report (March 2001), http://www.methanol.org/pdfFrame.cfm?pdf=Methanol_humantox_rev.pdf
28. M. Jamali, G.A. Lawrence, and K. Maloney, Fate of Methanol Spills into Rivers of Varying Geometry (September 2002), http://www.methanol.org/pdfFrame.cfm?pdf=MethanolSpillsMX.pdf
29. L. Smith, J. Molson, and K. Maloney, Potential Impacts on Groundwater of Pure-Phase Methanol Releases (September 2002), http://www.methanol.org/pdfFrame.cfm?pdf=GroundwaterMX.pdf
30. World Health Organization, International Program on Chemical Safety, Environmental Health Criteria 196, Methanol (1997), http://www.inchem.org/documents/ehc/ehc/ehc196.htm#SectionNumber:11.1
31. Molson, and K. Maloney, Potential Impacts on Groundwater of Pure-Phase Methanol Releases (September 2002), http://www.methanol.org/pdfFrame.cfm?pdf=GroundwaterMX.pdf

32. World Health Organization, International Program on Chemical Safety, Environmental Health Criteria 196, Methanol (1997), http://www.inchem.org/documents/ehc/ehc/ehc196.htm#SectionNumber:11.1
33. J. Brent, K. McMartin, S. Phillips, C. Aaron, and K. Kulig, "Fomepizole for the treatment of methanol poisoning," *New Engl. J. Med.*, 344, No. 6 (2001) pp. 424–429
34. M. Soffritti, F. Belpoggi, D.D. Esposti, L. Lambertini, E. Tibaldi, and A. Rigano, "First experimental demonstration of the multipotential carcinogenic effects of aspartame administered in the feed to Sprague-Dawley rats," *Environ. Health Perspect.*, 114, No. 3 (2006) pp. 379–385
35. Federal Register, Vol. 68, No. 73 (April 16, 2003) pp. 18743
36. World Health Organization, International Program on Chemical Safety, Health and Safety Guide No. 105, Methanol (1997), http://www.inchem.org/documents/hsg/hsg/v105hsg.htm#PartNumber:7
37. http://www.bitrex.com

Chapter 3

METHANOL REFORMING

Methanol is arguably the easiest organic molecule to chemically convert to a hydrogen-rich synthesis gas stream. This process, generically called reforming, requires an oxidant—usually either air or water, although hydrogen peroxide is also an option. Regardless of the choice of oxidant, the desired reaction pathway is oxidation of carbon to carbon dioxide with the liberation of hydrogen. If water is the oxidant, the process is called steam reforming. If air is the oxidant, the process is called partial oxidation. And if both water and air are oxidants, the process is called autothermal reforming.

For hydrogen fuel cell systems methanol has long been recognized as an attractive potential source of hydrogen. Indeed, methanol has been described as an ideal hydrogen storage medium, or hydrogen carrier, when one considers that methanol is made by reacting hydrogen with carbon monoxide and carbon dioxide, and then subsequently decomposed to the reactants.

3.1 Methanol Steam Reforming

At modest temperatures of 200°C to 400°C methanol readily reacts with steam in the presence of a catalyst to form a mixture of carbon oxides and hydrogen. Formation of excess by-product methane can

Methanol Fuel Cell Systems: Advancing Towards Commercialization
By Dave Edlund

be a problem at higher temperatures. Under ideal conditions the favored reaction stoichiometry is shown as equation 3.1.

$$CH_3OH_{(g)} + H_2O_{(g)} \rightarrow CO_2 + 3H_2 \ \Delta H°298 = +49.3 \text{ kJ/mol} \quad (3.1)$$
$$\Delta G°298 = -3.5 \text{ kJ/mol}$$

This ideal reaction stoichiometry illustrates one of the major advantages of steam reforming: ⅓ of the product hydrogen is derived from water. For this reason, and since water and methanol are miscible in all proportions, methanol-steam reforming is the most common approach to extracting hydrogen from methanol.

If the reactants are in the liquid phase, the heat of vaporization of methanol ($\Delta H_v = 37.6$ kJ/mol) and of water ($\Delta H_v = 44.0$ kJ/mol) increases the magnitude of the reaction enthalpy.

$$CH_3OH_{(l)} + H_2O_{(l)} \rightarrow CO_2 + 3H_2 \ \Delta H°298 = +131.4 \text{ kJ/mol} \quad (3.2)$$
$$\Delta G°298 = +9.3 \text{ kJ/mol}$$

The most widely accepted mechanism for methanol-steam reforming is the sequence of the two reactions shown below:

$$CH_3OH \rightarrow CO + 2H_2 \quad (3.3)$$

$$CO + H_2O \leftrightarrow CO_2 + H_2 \quad (3.4)$$

Equation 3.3 is the thermal decomposition of methanol, and is considered to be practically non-reversible under typical reforming conditions. On the other hand, equation 3.4, the water-gas shift reaction (WGS), is reversible and likely to be at least partially under equilibrium control. Thus, carbon monoxide is normally a significant by-product. Methane may also form in substantial concentrations at high temperature and low space velocity. A more realistic reaction equation is

$$CH_3OH + H_2O \rightarrow xCO_2 + yCO + zCH_4 + mH_2 + nH_2O \quad (3.5)$$

In equation 3.5 it is assumed that all methanol reacts with a stoichiometric amount of water and is consumed, and that the formation of methane results in production of by-product water according to equations 3.6 and 3.7.

$$CO + 3H_2 \rightarrow CH_4 + H_2O \quad (3.6)$$

$$CO_2 + 4H_2 \rightarrow CH_4 + 2H_2O \quad (3.7)$$

Because formation of methane represents a loss of 3 or 4 molar equivalents of hydrogen, it is usually not desirable to operate under conditions in which more than a fraction of a percent of methane forms. Thus, for practical purposes formation of methane can usually be ignored.

If we assume that by-product formation of methane is effectively suppressed, a more realistic steam-reforming reaction stoichiometry is shown in equation 3.8.

$$CH_3OH + H_2O \rightarrow 0.33CO_2 + 0.67CO + 2.33H_2 + 0.67H_2O \quad (3.8)$$

The relative ratio of CO and CO_2 reflected in equation 3.8 is typical for a platinum catalyst operating at 350°C to 400°C. The stoichiometry reflects complete conversion of methanol and only partial reaction of water. In fact, the conversion of water to desired products is largely governed by the water-gas shift reaction kinetics and equilibrium. Thus, the relative concentration of CO in the product reformate stream is expected to increase with increase in operating temperature assuming adequate reaction rate for the shift reaction. If the combination of catalytic activity for the shift reaction, reaction temperature, and space velocity is such that the overall rate of CO reaction with water is slow under the reaction conditions, then the concentration of CO may increase as temperature is decreased. An evaluation of CO concentration plotted against space velocity at constant temperature will indicate if reaction kinetics are limiting (if CO concentration increases with increasing space velocity at constant operating temperature, then the WGS reaction rate is limiting and true equilibrium control is not governing the CO concentration).

Referring to equation 3.8, it is clear that the concentration of hydrogen in the reformate product stream is relatively high at 58.25% by volume. If the product stream is dried, the concentration of hydrogen is even greater, 69.97% by volume. This high concentration of hydrogen is a benefit when considering pressure-driven hydrogen purification methods for purifying hydrogen from the reformate stream; for example, using a membrane process.

3.1.1 *System Design and Energy Balance*

Given that methanol and water are completely miscible and that steam reforming is accomplished at temperatures less than 450°C using a variety of commercially available catalysts, the typical system design is relatively simple. An example is the process flow diagram shown in Fig. 3.1. This system design assumes that the reformate stream from the steam reformer subsequently passes through a temperature-controlled single-stage shift reactor to increase the hydrogen yield. A two-stage shift reactor could easily be substituted for the single-stage reactor at an increase in cost. But if the steam reformer is operated at a relatively high temperature, e.g., greater than about 300°C, the added cost might be worthwhile. The orifice in the feed line just prior to the vaporizer/reformer yields a constant flow rate to the reformer, and is optional.

Since shift reactors alone cannot achieve suitably low CO concentration for use of the product reformate stream in low-temperature PEM fuel cells, the diagram indicates that the product stream is to be sent to a high-temperature PEM fuel cell (implying a CO concentration ≤3% by volume). The anode purge gas stream comprises the fuel gas stream. The purge ratio should be chosen to satisfy both

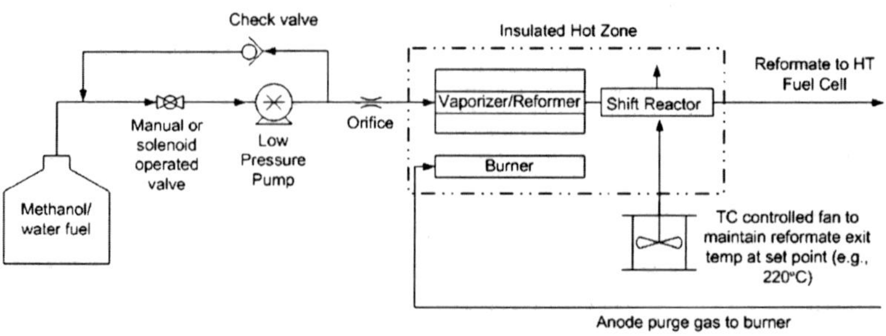

Figure 3.1 A simplified process flow diagram for a methanol steam reformer coupled with a water-gas shift reactor.

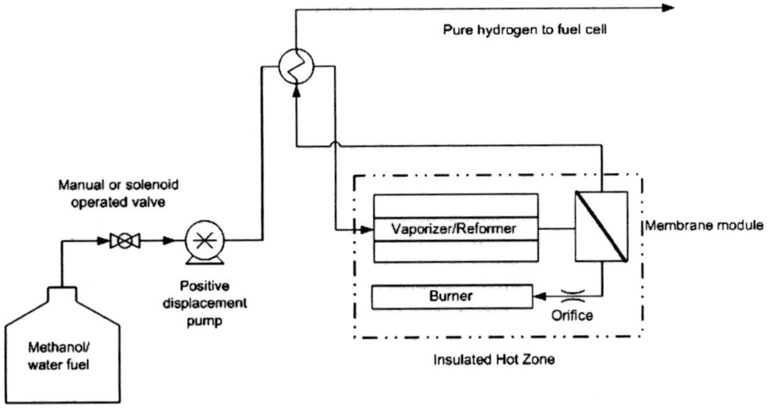

Figure 3.2 A simplified process flow diagram for a methanol steam reformer using a palladium-based hydrogen-purification membrane module.

acceptable kinetics at the anode side of the fuel cell and sufficient heating value to meet or exceed the enthalpy requirements of the steam reformer.

A different hydrogen purification means is required to produce hydrogen from methanol-steam reforming that meets or exceeds purity requirements for low-temperate PEM fuel cells. Hydrogen purification is addressed in more detail in Chapter 4. However, as a comparison to Fig. 3.1 a process flow diagram for a methanol-steam reformer using a membrane-based purifier is shown in Fig. 3.2. Very high purity hydrogen can be achieved from reformate in a single stage, thus allowing integration of the reformer with low-temperature PEM fuel cells.

Aside from the differences in purification technology, the two systems are remarkably similar. The need for elevated pressure for membrane-based hydrogen purification places specific requirements on the methanol/water feed pump that are not otherwise necessary. However, increasing pressure has benefits to the steam reformer including a reduction in space velocity at the same mass flow rate. Since the thermal decomposition of methanol (equation 3.3) is not reversible under typical reforming conditions, there is not significant impact of pressure on conversion of methanol and water to products. And the WGS reaction is not influenced by

Table 3.1 Summary of heat input requirements for a methanol-steam reformer based on the reaction stoichiometry in equation 3.8

	Heat Input (kJ/gmol Methanol)	
	Without Thermal Recovery	With Thermal Recovery
Heating liquid methanol/water	8.82	8.82
Vaporization	81.90	81.90
Heating methanol/water vapor	21.17	21.17
Reaction	76.75	76.75
Heating combustion air	128.28	4.77
Heat recovery, reformate	0	-(25.20)
Total	316.92	168.21

pressure since the reaction is not accompanied by a change in molar volume.

Both process flow diagrams imply a vaporizer that is integral with the catalytic steam reformer. It makes sense to closely couple these two unit operations to minimize heat loss from any connecting tubes or piping. However, to analyze the energy input requirements, the vaporizer and reforming reactor will be considered separately (Table 3.1). Heat recovery can be employed to reduce the required heat input to the reformer system. The decision to add heat exchangers for thermal recovery must be based on an assessment of relevant system impact (e.g., size and weight) as well as system cost.

Two cases are compared in the table: one assumes no thermal recovery and the other assumes thermal recovery from product reformate and combustion exhaust (no hydrogen purification; process flow diagram similar to Fig. 3.1). Thermal energy recovered from the hot product reformate stream could be transferred to the incoming methanol/water feed stream, whereas thermal energy from the combustion exhaust could be transferred to the incoming combustion air (Fig. 3.3). Where thermal recovery is considered, a rather modest approach was adopted; no credit is taken for the latent heat of water vapor in the reformate stream and combustion exhaust stream. It is doubtful that the concentration of water vapor in the combustion exhaust is high enough to warrant any serious effort to

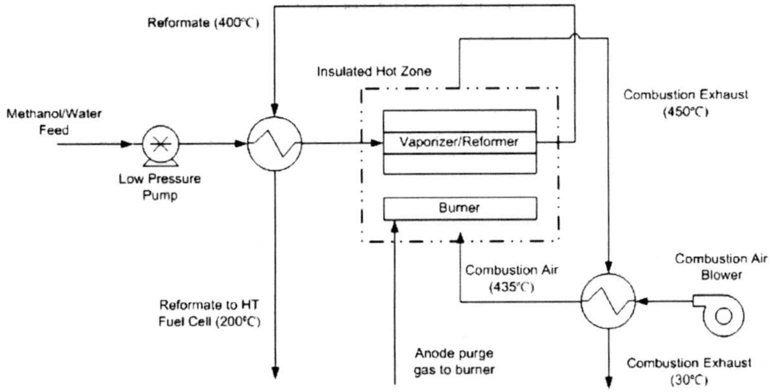

Figure 3.3 Process flow diagram for a methanol-steam reformer serving as the basis for the enthalpy analysis in Table 3.1.

recover its heat value. However, the concentration of water vapor in reformate may be sufficiently high, depending on the methanol-to-water fuel mixture and operating conditions of the reforming reactor, to make it worthwhile to recover some of the latent heat from the water vapor.

In calculating the enthalpy values in the table it was assumed that the methanol/water feed stream is initially at 15°C and is a 1:1 molar ratio (64 wt% methanol). The reforming reaction is assumed to take placed at 400°C resulting in 100% conversion of methanol and the product reformate stream is at 400°C. The combustion air, 3 times stoichiometric excess, is delivered to the burner at 15°C and is heated to a combustion temperature of 700°C, transferring a significant portion of its energy to the vaporizer/reactor and exiting the reformer at 450°C. The product reformate composition is assumed to be 8.25% CO_2, 16.75% CO, 58.25% H_2, and 16.75% H_2O (all on a volume basis, reference equation 3.8). Thus, one mole of methanol plus one mole of water is reformed to yield 2.33 moles hydrogen.

Clearly the thermal energy required to heat the combustion air is the single greatest variable in the overall thermal energy balance. Without heat recovery, the thermal energy necessary to heat the combustion air exceeds 34% of the total thermal energy requirement. However, if the reformate stream is cooled from 400°C to 200°C by heat exchange (for instance with the incoming methanol/water stream), then the amount of thermal energy recovered is 25.20 kJ/gmol methanol; this thermal energy is transferred into the methanol/water feed stream. More significantly, as much as 93% of the thermal energy in the 450°C combustion exhaust exiting the reformer can be recovered and potentially transferred to the incoming combustion air stream. Doing so accounts for nearly 100 kJ/gmol methanol as recovered thermal energy.

What is not obvious from Table 3.1 is that recovered thermal energy indirectly reduces the energy requirements of the steam reformer by reducing the amount of fuel that must be burned and correspondingly, reducing the amount of combustion air that must be heated.

3.1.1.1 Energy efficiency

The preceding summary of enthalpy requirements for steam-reforming methanol can also serve as a basis for understanding the energy efficiency of steam reforming. The thermal efficiency of a reformer is defined as

$$\eta_{\text{LHVSR}} = \frac{\left(RERORM_{\text{th}} - FUEL_{\text{th}}\right)}{MeOH_{\text{th}}} \tag{3.9}$$

where η_{LHVSR} is the thermal energy efficiency (lower-heating value) of the steam reformer, $REFORM_{\text{th}}$ is the lower-heating value of the product reformate stream, $FUEL_{\text{th}}$ is the lower-heating value of the fuel stream to the burner, and $MeOH_{\text{th}}$ is the lower-heating value of the methanol fraction of the feed stream. In the examples presented in this chapter the only gases in the product reformate stream that have fuel value are H_2 and CO, and the latter can be converted to H_2 via the water-gas shift reaction. Based on the reaction stoichiometry shown in equation 3.8 and the enthalpy requirements shown in Table 3.1, the following values can be assigned to the variables:

$$REFORM_{\text{th}} = 753.47 \text{ kJ/gmol methanol}$$

$$FUEL_{\text{th}} = 316.92 \text{ kJ/gmol methanol (without thermal recovery)}$$

$$MeOH_{\text{th}} = 676.1 \text{ kJ/gmol methanol}$$

Thus, $\eta_{\text{LHVSR}} = \dfrac{(753.47 - 316.92)}{676.1}$ kg/gmol methanol, or 65% efficiency without heat exchangers for thermal recovery. If we consider the case where thermal recovery is employed,

$$FUEL_{\text{th}} = 168.21 \text{ kJ/gmol methanol (with thermal recovery)}$$

and $\eta_{\text{LHVSR}} = \dfrac{(753.47 - 168.21)}{676.1}$ kJ/gmol methanol, or 87% efficiency with thermal recovery.

This energy efficiency calculation illustrates an interesting point: the thermal efficiency of a steam reformer can exceed 100%. This non-obvious result is due to the fact that hydrogen (which has considerable heating value) is a product derived partly from the reaction of water (water has no heating value). In the limiting case in which the steam reforming of methanol proceeds according to the ideal reaction stoichiometry (equation 3.1) and we imagine this reaction going to completion without the input of any fuel value, then the thermal efficiency we would calculate is 107% (LHV) and 126% (HHV). In the steam-reforming reaction, the energy that is required to reduce water to hydrogen results mostly from the oxidation of the carbon in methanol to CO_2 with a smaller contribution from the increase in total entropy.

3.1.1.2 Methanol-reforming catalysts

Catalysts for methanol-steam reforming can be dividing into two groups based on chemical composition: base-metal formulations and platinum-group metal (PGM) formulations. Historically, base-metal commercial WGS catalysts have been used very effectively. Low-temperature WGS catalysts and methanol synthesis catalysts based on a copper/zinc oxide/alumina formulation (e.g., BASF K3-110) exhibit very high activity but suffer from rapid degradation at temperatures >280°C due to copper sintering. The catalyst, which is active in the reduced-copper oxidation state, is extremely pyrophoric. High-temperature WGS catalysts incorporating copper, chromium and/or iron, and zinc oxide also are active for methanol-steam reforming. But again, these base-metal formulations only have appreciable catalytic activity in the reduced oxidation state and they are pyrophoric.

Base-metal catalyst formulations are usually offered as compressed, homogeneous pellets. It is not standard practice to apply the base-metal catalyst formulation as a wash-coat onto a porous substrate. Consequently, these formulations are used in packed-bed catalytic reactors, and particle physical attrition due to thermal cycling, shock, and vibration can be a serious problem.

PGM catalyst formulations are usually based on either platinum or palladium; both are active for methanol-steam reforming. Typically the PGM is dispersed onto a porous alumina substrate due to the high cost of platinum and palladium. The porous substrate may be in a wide range of physical forms including pellets and other geometric shapes, and as a wash-coating on a structured support (such as corrugated foils and foams).

It is generally accepted that platinum has higher specific activity than does palladium. In addition, platinum is suggested to be more resistant to thermal sintering than is palladium. Süd-Chemie offers a platinum on alumina catalyst designated FCRM-9B that is active for methanol-steam reforming and exhibits stable performance for more than 1,000 hours when operated at 400°C. FCRM-9B can be purchased in a granular (extrudate) form or as a wash-coat applied to an engineered support.

The CO concentration in the reformate stream is dependent on several operational variables including temperature, space velocity, and methanol-to-water ratio. At 400°C and using a 1:1.04 methanol/water mix (slight excess of water), the CO concentration in the reformate stream will be about 5% at a space velocity* of about 3,000 hr^{-1}, and about 20% at a space velocity of 22,000 hr^{-1}. If under equilibrium control, increasing the operating temperature will result in an increase in CO concentration. Decreasing the methanol-to-water ratio will typically result in a reduced CO concentration in the product reformate stream.

Although platinum on alumina is very active, demonstrating complete conversion of methanol at 400°C and 22,000 hr^{-1} GHSV,

* GHSV is calculated as liters of dry reformate generated per hour divided by the volume of the catalytic reactor; no correction is made for the volume occupied by the catalyst.

addition of ceria is reported to further increase activity. Pure methanol is reported to be completely decomposed to CO and H_2 over a platinum/ceria/alumina catalyst (5% platinum and 10% ceria loading on alumina) at 315°C and a 42,150 hr^{-1} GHSV.[1] Since thermal decomposition of methanol is the first step in methanol-steam reforming, this is a good indicator of steam-reforming activity.

Palladium, when alloyed with zinc, is also a very active catalyst for methanol-steam reforming. However, unlike platinum, palladium-zinc catalysts demonstrate high selectivity for CO_2 over CO. This is very beneficial since CO is not only an anode poison in low-temperature PEMFCs, but CO is equivalent to H_2 through the WGS reaction. Thus, a reduction in by-product CO represents an increase in H_2 yield from the methanol/water feedstock. Palladium-zinc catalysts are typically prepared on an oxide support, usually from a palladium precursor, zinc oxide, and the oxide support material. BASF offers a commercial palladium-zinc methanol-steam reforming catalyst designated RM-05.

Employing RM-05 at 400°C and using a 1:1.04 methanol/water mix, the CO concentration in the reformate stream will be about 2–3% at a space velocity of 22,000 hr^{-1}. Lowering the operating temperature increases the selectivity to CO_2.[2]

Using platinum catalysts (and Group VIII metals in general), the general mechanism for steam-reforming methanol is accepted as the sequential reactions shown as equations 3.3 and 3.4. The first step, methanol thermal decomposition, is thought to proceed via formation of intermediate formaldehyde; followed by rapid decarbonylation to CO and H_2. In contrast, the high selectivity of the palladium-zinc catalyst for CO_2 is thought to be due to the rapid reaction of water with intermediate formaldehyde resulting in formate, which subsequently decomposes to CO_2 and H_2.[3]

When selecting a catalyst for steam-reforming methanol, it is important to give consideration to the product design, scale, operating mode, and target cost. Base-metal catalysts will be favored where the higher cost of a PGM catalyst cannot be tolerated. However, the pyrophoric nature of base-metal catalysts may be considered an unacceptable safety risk for some applications. Likewise, shock and vibration tolerance may urge the design engineer to select a wash-coated catalyst, in which case PGM selections are likely to dominate.

3.1.2 *Reactor Designs*

Since steam reforming of methanol (or, for that matter, other alcohols and hydrocarbons) is a strongly endothermic reaction sequence, the observed rate of reaction is often limited by the rate of heat transfer to the catalytic reaction sites. For example, conventional tubular packed-bed catalytic reactions will exhibit large radial temperature gradients between the reactor wall and the central portion of the bed.

The simple experimental apparatus depicted in Fig. 3.4 can be used to measure the radial temperature gradients in experimental reactors. An external heat source (e.g., electric furnace) is used to heat the vaporizer and the catalytic reformer. By encasing the reformer in an aluminum block, the skin temperature can be regulated to a uniform and controlled temperature using a thermocouple inserted into the aluminum block (TC 1) and providing feedback control to the furnace. A second thermocouple (TC 2) inserted into the catalytic reforming reactor along the central axis measures the axial temperature.

The aluminum block must be intimately in contact with the wall of the reactor. Two suitable methods of fabricating the reactor with aluminum block include casting aluminum around the reactor (without catalyst) and clamping two halves of an aluminum block through which a cylindrical hole has been bored to fit tight against the

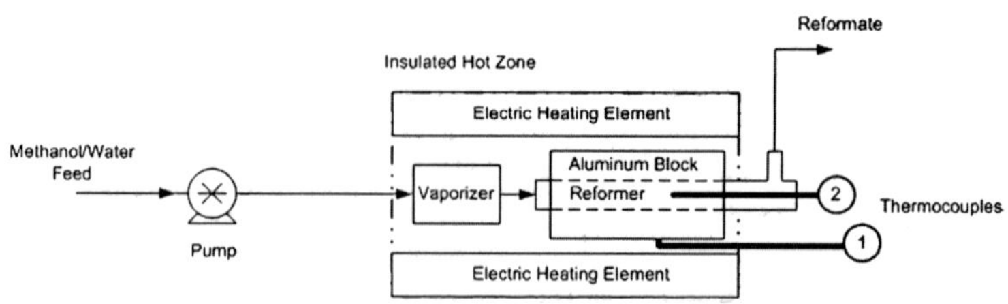

Figure 3.4 Experimental apparatus for measuring radial temperature gradients in a tubular reforming reactor.

reactor wall. The high thermal conductivity of aluminum ensures that there are no hot spots or cool spots on the reactor wall.

In one experiment using a stainless steel tubular reactor (0.625-inch outside diameter) containing 22.8 grams of FCS-12 extrudate (a platinum catalyst offered by Süd-Chemie, Louisville, Kentucky, USA) the wall temperature of the reactor was controlled to 395°C and the axial temperature was measured to be 279°C. For this experiment the feedstock was 61.3 wt% methanol in water and the feed rate was 2.5 g/min.

Thus, although the temperature of the aluminum block and the wall of the reactor were controlled to about 400°C, the actual axial temperature within the reactor was 116°C lower. Such a large decrease in temperature within the reactor will result in a significant decrease in observed reaction rate and influence the water-gas shift equilibrium.

Indeed, reforming reactor design is really a combined exercise in heat transfer and mass transfer. Two very simple, yet powerful, design guidelines are as follows:

- Ensure thorough contact between the feed stream and catalytic sites; avoid channeling.
- Facilitate rapid heat transfer; heat transfer should not be rate-limiting.

These guidelines apply equally well to conventional packed-bed reactors, microchannel reactors, and reactors based on an engineered catalyst substrate.

3.1.2.1 Conventional packed-bed reactors

Historically very important, packed-bed reactors employing a granular extruded or pelletted form of the catalyst are still extremely versatile and are perhaps the lowest cost of all reactor configurations. A multitude of textbooks have been devoted to the design of packed-bed reactors so as to ensure good mass transfer and avoid channeling. Assuming a tubular or cylindrical reactor shell, a robust design guideline is to use a catalyst particle size that is no more than 0.2 times the reactor diameter and the reactor length should be at least 5 times the diameter (10 times the diameter is even better).

Fortunately, catalysts are manufactured in a multitude of shapes and sizes so that properly sizing the catalyst for the reactor dimensions should not be a problem.

As mentioned above, to avoid large thermal gradients between the tube or shell wall and the interior of the catalyst bed, the reactor diameter should be small—0.5 inches or less. To achieve adequate volume for the required amount of catalyst, multiple tubes may be arranged in parallel by welding the tube ends to a sheet header; analogous to a shell-and-tube heat exchanger. This highlights a drawback of tubular packed-bed reactors: if a relatively large amount of catalyst is required (i.e., large reactor volume), then optimizing the design for fast heat transfer—high surface-area-to-volume ratio—implies a more complex and expensive reactor design.

A time-proven means to enhance the rate of heat transfer with tubular reactors is to braze metal fins onto the outer wall of the reactor tube. So-called finned-tubes are commercially available wherein the tube and fins may be stainless steel, carbon steel, aluminum, copper, or copper alloys. The metal selected for the fins may, in some cases, differ from the metal the tube is made from. For instance, aluminum fins may be brazed onto the outer surface of a stainless steel tube. A few manufacturers also offer integral-finned tubes, but the metal selection is limited to copper alloys and carbon steel. Also, an option available with integral-finned tubes is internal ribs to induce turbulence along the interior wall of the tube. The addition of fins to the outer surface of a tube will greatly increase the available heat-transfer surface area; 10 times or more area increase is possible.

A limitation with conventional heating of reactors (tubular, annular, or other shape) is the stagnant gas boundary layer between the hot combustion gas and the wall of the reactor. A method to reduce the impact of this boundary layer is to encase the reactor wall in a thermally conductive metal such as aluminum or copper. Aluminum tolerates high temperatures better than copper, and aluminum is also less expensive and only 30% the density of copper. Casting a block of aluminum around the reactor is easily done by a foundry, although if the reactor is tubular, machined blocks may also be used to bring the aluminum mass into contact with the reactor wall.

Since metals, especially aluminum and copper, have very high thermal conduction relative to gases, the effect of a gas-phase boundary layer on heat transfer into the reactor may be minimized. Of course, there is still the issue of heat transfer from the hot combustion gases into the aluminum block, but the surface area and features of the aluminum block can be optimized for heat transfer without concern over reactor geometry. Thus the two challenges of reactor design and heat transfer may be largely separated.

Packed-bed reactors are by no means geometrically limited to cylindrical tubes. Annular reactors are also very useful and, in fact, can yield a reactor with large catalyst volume and high surface-area-to-volume ratio in a design that is more compact and substantial less expensive than multiple tubes welded to headers. The challenge with annular reactors is ensuring adequately uniform flow distribution through the bed. To understand this challenge, it is helpful to conceptualize an annular reactor as a circular arrangement of close-spaced tubes fed from a common manifold.

The versatility of this type of reactor stems from the ease of assembly and loading with catalyst, plus the fact that any catalyst formulation can be made in a suitable shape and poured into the reactor. The commercial selection of catalysts is diverse, and less expensive than formulations that must be applied to a substrate in a multi-step process.

Perhaps the most significant disadvantage of packed-bed reactors is catalyst attrition due to thermal cycling, and mechanical shock and vibration. Settling of catalyst beds, resulting in channeling, is easily avoided by orienting the bed vertically rather than horizontally. Attrition may be managed by segmenting the bed and/or installing one or more high-surface-area filters downstream of the bed. Such filters (e.g., stainless steel filter cloth) may even be installed within the catalyst bed.

3.1.2.2 Microchannel reactors

Microchannel reactors are designed to maximize the surface-to-volume ratio and to utilize short thermal conduction pathways.[4] The result is a very large increase in observed reaction rate with typical space velocities that may be 5 to 10 times what can be achieved in a packed-bed reactor. Microchannel reactors are most commonly made from metal by appropriate micro-machining techniques including chemical etching. A block of metal may be shaped or the reactor may be assembled from a stack of thin metal plates, brazed or diffusion bonded together, to yield a monolithic block with intricate internal flow channels.

Much of the early reported work on micro-channel reactors, including fabrication methods and designs, was published by Pacific Northwest National Laboratory (Richland, Washington, USA). Some commercial rights were later licensed to Velocys (Plain City, Ohio, USA). For purposes of patent definitions, Pacific Northwest National Laboratory has defined a microchannel reactor as having fluid flow channels ≤1 mm in cross section. It is not unusual for the reactant flow channels to be on the order of 0.1 mm in cross section. This facilitates excellent mass transfer rates to the catalyst surfaces.

The small dimensions of microchannel reactors may make them susceptible to clogging due to non-volatile residue dissolved in methanol and water and carried into the reactor. For example, the IMPCA specification for methanol requires that non-volatile residue be ≤8 mg/L. Assuming the feed rate of methanol into a reformer is 50 mL/min (appropriate for the production of about 35 NLm product hydrogen), over an operational lifetime of 500 hours the total amount of methanol flowing through the vaporizer and reactor will be 1,500 L. If the methanol contains only 4 mg non-volatile residue per L (50% of the maximum allowable), the total weight of residue entering the reformer will be 6 g, which could easily result in blocked flow channels.

The catalyst must be applied to the wall of the microchannel reactor both to realize the benefit of extremely fast heat transfer as well as to manage pressure drop through the reactor. Much has been published on applying wash coatings and active catalysts to metal surfaces, and this applies well to microchannel reactors. Copper-zinc oxide,[5] platinum, palladium, and palladium-zinc formulation are popular for methanol reforming.[6,7,8]

In addition to exhibiting very fast reaction rates, microchannel reactors are robust with respect to catalyst attrition assuming the catalyst is formulated and applied using a method that yields strong adhesion with the underlying support metal (usually stainless steel).

Provided the catalyst does not flake off the support upon thermal cycling, catalyst loss is virtually non-existent. Thus, this type of reactor is favored for applications where space is minimal and shock and vibration are expected.

The two major drawbacks of microchannel reactors are relatively high weight and high cost. Since the reactor is a large block of metal with tiny flow passages, the weight will be significantly more than other types of reactors. The preferred material for making microchannel reformers is stainless steel. Thus, the reactors are commonly relatively heavy simply due to the mass of steel. Aluminum alloys, copper, and copper alloys are all unsuitable due to relatively low mechanical strength at the elevated operating temperature of the reformer.

The favored method for fabricating microchannel reactors—bonding together a stack of thin machined plates—is an expensive approach. Certainly there are many applications were weight is not a strong driver, but cost is almost always an important, if not dominant, consideration. Until lower-cost fabrication methods are developed, microchannel reactors are not likely to be widely commercialized for reformed methanol fuel cell systems.

3.1.2.3 Reactors using engineered catalyst supports

As commonly used, the term "engineered catalyst support" refers to a foam, corrugated metal foil, monolith, etc., that is thermally stable (with respect to chemical and physical changes) and either exactly or closely approximates a single unit rather than a collection of catalyst pellets or shapes (e.g., wagon wheels, rings, cylinders, spheres, extrudates). An excellent example of an engineered catalyst support is the cordierite honeycomb monolith that is used in most automobile catalytic converters. The foam, corrugated metal foil, monolith, etc., shape is inserted into the reactor shell. Catalyst is applied to the support structure either prior to insertion into the reactor shell or after insertion into the shell.

Engineered catalyst supports are widely used in automotive applications (catalytic converters) since conventional granular or pellet catalysts would break-down under the shock and vibration associated with transportation. The microchannel reactor discussed

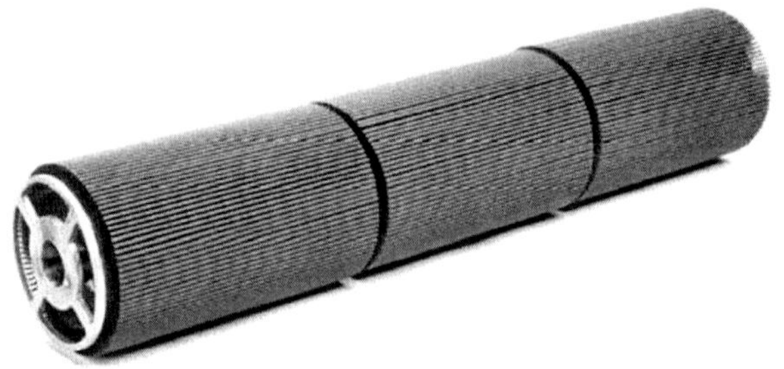

Figure 3.5 Example of an engineered catalyst support for an annular reactor. The support is a folded (corrugated) sheet metal that has been coated with catalyst (photo courtesy of Catacel Corp.). See also Color Insert.

above is another example of an engineered catalyst support; however, in that case rapid heat transfer is also incorporated into the design of the catalyst support. The engineered catalyst support structure is integral with the wall or body of the reactor. Another example of an engineered catalyst support is the annular reactor design shown in Fig. 3.5.

The pros and cons of conventional catalyst shapes used in packed-bed reactors and engineered catalyst supports as applied to methanol reforming are compared in Table 3.2.

Table 3.2 Comparison of methanol reforming reactors using a packed catalyst bed and an engineered support with applied catalyst

	Advantages	Disadvantages
Packed Bed	• Low cost • Flexible (easy to change catalyst) • Wide selection of commercial catalysts • Amenable to rapid heat-transfer reactor designs	• Subject to catalyst attrition (amplified by shock, vibration, thermal cycling) • Settling and channeling can occur with improper reactor design
Engineered Support	• Durable to shock, vibration, thermal cycling • Amenable to rapid heat-transfer reactor designs	• Higher cost • Not flexible • Limited selection of commercial catalysts

The engineered catalyst support may be made of metal or ceramic; metals offer the advantage of better thermal conduction but ceramics offer the advantage of better thermal stability. Other considerations for choosing or designing the support include surface-area-to-volume ratio; pressure drop; cost; and durability (adhesion of the catalyst coating to the support material).

Corrugated stainless steel foil has been long used as an engineered catalyst support. When rolled into a coil, the result is a compact metal structure with straight-through flow channels and reasonably good thermal conduction through the metal corrugations from one layer to the next. This structure is also very inexpensive to manufacture. The straight-through flow channels aid application of the catalyst by wash-coating processes, and result in very low pressure drop. Such engineered catalyst supports are used for emission-reduction applications.

The primary advantage of engineered catalyst supports is that there are no catalyst pellets to rub together; rather the entire structure is immobile. Shock, vibration, and thermal cycling can cause conventional granular or pelletized catalysts in packed-bed reactors to abrade and physically deteriorate; catalyst dust is usually carried downstream where it may plug valves and other orifices. Also, catalyst is lost from the reactor leading to a decline in throughput over time. Catalyst that has been coated onto an engineered support, thereby becoming immobile, will not suffer from this degradation mechanism. This is why ceramic monoliths and stainless steel corrugated rolls are used as catalyst supports for vehicular emission-reduction applications.

But the drawback to engineered catalyst supports is that they are often more expensive than are traditional catalyst pellets or other shapes. This is especially true for metal and ceramic foams. Not only is the substrate expensive, but a secondary process is required to apply the catalyst. Depending on the composition of the support structure, an alumina wash coat, with or without other additives, is applied as a slurry to the support; more than one application of the wash-coat slurry is frequently necessary to obtain adequate coverage. The wash coat is allowed to dry and then it may be calcined. To achieve good adhesion between the wash coat and the underlying support material, it may be necessary to clean and oxidize the support surface prior to applying the wash coat (for instance, if the support is a metal, such as stainless steel). After the alumina wash coat is calcined, then the precatalyst is applied usually by a wet impregnation process, followed again by calcining.

The specific physical features of the engineered support are also important—if the channels are tortuous and/or very small in cross-section, the wash-coat application will plug some (many) of the channels and increase pressure drop as well as decrease the amount of catalyst surface area per unit volume.

Exactly how significant the cost difference is between a conventional packed-bed reactor and one utilizing an engineered support has to be calculated in the context of scale and application. The cost of a catalytic reactor does not increase linearly with scale (output); it increases at a slower rate, usually following a power function where the exponent is <1. Thus, one would expect small-scale reactors to be more sensitive to cost than larger reactors. In other words, small portable and back-up fuel-cell systems using a methanol reformer will suffer greater cost pressure than will larger stationary, vehicular, and back-up systems.

3.1.2.3 Trade-offs: cost vs. functionality, sizing, durability

Considering the many different options available, it may seem a daunting task to rationally design an optimal methanol-reforming reactor for fuel-cell applications. But with careful consideration to the market requirements (technical performance, durability, price), the field of options can be reduced and a clear direction selected.

A few simple guidelines can be very helpful:

- If the application is for use on a mobile platform (man-portable, vehicle mounted, etc.), then an engineered catalyst support may the better choice over a packed-bed reactor to avoid catalyst attrition and bed settling than can lead to channeling.
- Intermittent-use applications will generally require an operational lifetime of 500 hours to 3,000 hours whereas continuous-use applications will require operational lifetimes >10,000 hours.
- Continuous operation and stationary applications should favor packed-bed reactors.

- If weight is a primary driver, then high heat-transfer rates (microchannel designs) should be carefully evaluated.
- Low-power fuel-cell applications (requiring smaller reformers) will be subject to greater cost pressure than will larger reformers.

Example 1. Experience at Protonex Technology Corp developing a 250 W scale reformed-methanol PEMFC illustrated the importance of scale on cost differential. The reformer was only required to deliver 4 NLm of purified product hydrogen to the fuel cell. To achieve this, approximately 9 to 10 NLm of methanol reformate was necessary (a palladium-alloy membrane purifier was used to yield pure product hydrogen from the reformate stream).

Using an aluminum foam engineered support manufactured by a US corporation, the cost of the foam plus catalyst coating (quoted by two commercial vendors) was about \$450 each in quantity <100 (the aluminum foam represents about 25% of this total cost). At higher volume, about 1,000 units, the cost of the aluminum foam and catalyst coating was less, approximately \$300 each. Even in high volume (>5,000 units annually) the projected cost did not fall below \$220 each. A significant cost driver was the labor required to obtain a sufficient coating of the catalyst layer, since multiple applications of the alumina wash-coat were required.

In comparison, an extruded catalyst (1.6 mm extrudate) of the same formulation as used with the aluminum foam cost only \$100 for <100 reactors annually. In high volume (>5,000 reactors annually) the cost of catalyst was projected to be only \$35.

Example 2. A kilowatt-scale methanol reformer was developed at IdaTech, LLC, to deliver up to 20 NLm of pure hydrogen (again, using a palladium-alloy membrane purifier). This methanol reformer was intended for stationary application where shock and vibration were not important considerations. In addition, weight was not an important factor. For this reformer, cost was the dominant consideration and consequently a base-metal catalyst in the form of ⅛-inch pellets poured into a packed bed reactor was the best choice.

To minimize the effect of catalyst settling and channeling, the reactor was oriented in a vertical arrangement. A particulate filter downstream from the reformer, and upstream of the membrane purifier, prevented catalyst dust from plugging small flow channels and orifices.

This design provided durable performance of several thousand hours (up to 10,000 operating hours was demonstrated) despite the fact that packed-bed reactors are commonly faulted for poor durability. Success is attributed to the vertical orientation of the reactor, thus prevented channeling as the catalyst bed settles; oversizing of the catalyst bed (acceptable due to the relatively low cost of the base-metal pelletized catalyst); and the target stationary application that did not involve substantial shock and/or vibration.

3.2 Partial Oxidation and Autothermal Reforming

Although methanol-steam reforming is the dominant approach to producing hydrogen from methanol, two other primary methods are also available: catalytic partial oxidation (CPOx) and autothermal reforming (ATR). As the name implies, CPOx uses air as the oxidant whereas both air and water are oxidants in ATR—in fact, ATR is a combination of CPOx and steam reforming. The two methods also differ in the enthalpy of reaction: CPOx is strongly exothermic while ATR is operated as close to zero net enthalpy as possible. The ideal net chemical reactions are shown below assuming that air is the source of oxygen:

$$CH_3OH + 0.5O_2 + 2.0N_2 \rightarrow CO_2 + 2H_2 + 2.0N_2 \quad (3.10)$$

$$\Delta H°298 = -192.5 \text{ kJ/mol}$$

$$CH_3OH + 0.1O_2 + 0.4N_2 + 0.8H_2O \rightarrow CO_2 + 2.8H_2 + 0.4N_2 \quad (3.11)$$

$$\Delta H°298 \approx 0.0 \text{ kJ/mol}$$

The CPOx ideal reaction is shown in equation 3.10. Unfortunately, such a high degree of selectivity for hydrogen is not attained in practice as a portion of the product hydrogen is invariably oxidized to water. If water is intentionally added to the feed stream in sufficient quantity to offset the exothermic CPOx reaction (according to equation 3.1 the steam-reforming reaction is endothermic by

49.3 kJ/mol) then we arrive with the ideal ATR chemical reaction as shown in equation 3.11. In deriving the ATR reaction stoichiometry no attempt was made to account for the heat capacity of the feedstock or the heat of vaporization of water; thus in practice these factors plus normal heat loss would argue for greater addition of oxygen (and correspondingly, nitrogen).

Recall that steam-reforming methanol ideally yields three moles of hydrogen for each mole of methanol that reacts. The yield of hydrogen from the CPOx reaction is substantially less. For this reason, combined with the dilution effect of nitrogen from air, the concentration of hydrogen in the product reformate stream at best is only 40% compared with 75% hydrogen in reformate from steam reforming. Consequently, CPOx has limited applicability in fuel cell systems: it is not adequately energy efficient.

Nevertheless, the fact that CPOx is exothermic and can start very quickly from a cold or ambient condition does have merit. Faced with the challenge of rapid reformer start-up for automobile applications, Xcellsis GmbH devised and demonstrated a hybrid methanol reformer subsystem that utilized a small CPOx reactor at start-up to provide both hydrogen-rich reformate for the PEMFC stack and heat to bring the primary steam reformer to operating temperature.

The ATR reaction is much more similar to the steam-reforming reaction than is CPOx as is evident by comparing equation 3.11 with equation 3.1. However, neither the steam-reforming reaction nor the CPOx reaction proceed as shown in equations 3.10 and 3.1, thus the ATR reaction shown above (equation 3.11) is unrealistically simplified. When thinking of the chemical reactions involved, this simplistic interpretation is useful, but it fails to provide an adequate interpretation of the energy balance and energy efficiency of the ATR process (see below).

Although the goal when operating a catalytic ATR reactor is to maintain very close to neutral enthalpy (heat being neither consumed nor released), in practice this is very difficult to achieve and maintain for any prolonged period of time. And since the ATR is not designed for the application of heat external to the reactor, one usually chooses to operate the reactor just slightly exothermic.

This naturally leads to another constructive way to differentiate ATR reactors from steam reformers: ATRs are internally heated and steam reformers are externally heated. It follows, then, that ATRs typically are heated from ambient temperature to operating temperature faster than are steam reformers; although the time difference is not as great as one might think because ATRs operate at a higher temperature than do steam reformers. As previously mentioned, a well designed and reasonably light-weight steam reformer may require between 15 minutes and 30 minutes for cold start-up, whereas an ATR may require 15 minutes to 20 minutes for cold start-up. If downstream hydrogen purification is employed, then additional time is needed to achieve full operational readiness.

3.2.1 *System Design and Energy Balance*

The ATR and CPOx system designs are very similar, differing primarily in that supplemental water (distilled or de-ionized) is added in the case of ATR. The reactor system designs are shown in Fig. 3.6. The CPOx and ATR reactor is usually relatively small and consists of a catalyst monolith (catalyst coated onto a honeycomb cordierite structure or foam), but it may also consist of a series of catalyst coated screens stacked atop one another.[9]

The ATR and CPOx reactor are often relatively small and they operate at high space velocity—gas-hourly space velocity as high

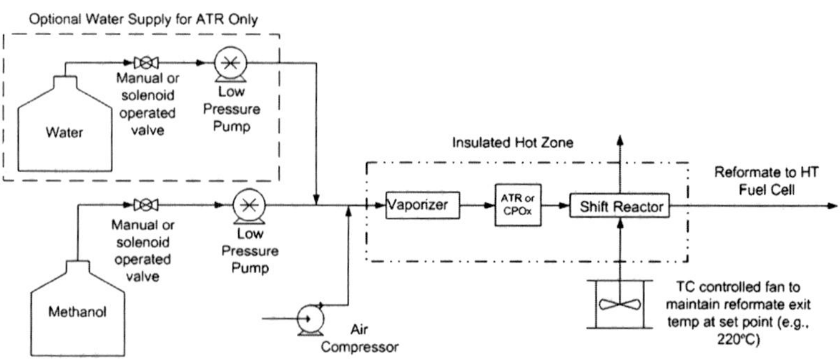

Figure 3.6 A simplified process flow diagram for methanol CPOx reactor and ATR coupled with a shift reactor. Optional water addition loop for ATR is shown in the upper right.

as 440,000 hr^{-1} has been reported for a methanol CPOx reactor using stacked catalyst-coated screens.[9] In contrast to methanol-steam reforming, the balance-of-plant is generally more complex for CPOx and ATR. This is due to the requirement to control the ratio of oxygen to carbon (air to methanol) and, if ATR is being carried out, the water-to-carbon ratio. Thus, flow measurement combined with a feedback control loop, and liquid pumps and air compressor (or blower) that can be turned up and down are required.

Typical oxygen-to-carbon ratios (O:C) for methanol CPOx should not exceed 1:1, and best results may be obtained at slightly lower ratios. If water is added for ATR, the preferred O:C ratio is reduced to 0.15:1[10] to 0.6:1[11] while the steam-to-carbon ratio is preferred to be about 1:1 to 1.5:1. Selectivity to hydrogen is very sensitive to the O:C ratio, excessive oxygen will result in oxidation of hydrogen to water. However, the hydrogen selectivity is not especially sensitive to the steam-to-carbon ratio. If water is added (ATR) it is important that the water be rather pure as mineral deposits may cause rapid decline in catalyst activity as well as cause scaling in the vaporizer.

The process diagram in Fig. 3.5 includes a vaporizer. This makes it easier to admit a premixed feed stream comprising methanol, air, and (optionally) water. Another option is to co-inject air and methanol (and water) directly into the CPOx reactor (or ATR if water is added). In this case an atomizing injector is used to cause the liquid to disperse into an aerosol as it enters the reactor. The aerosol droplets ideally vaporize immediately in front of the catalyst core.

The advantages of a vaporizer include lower cost and a long mixing region prior to the reactor. A vaporizer can also deliver expected performance over a broad range of throughput. The drawbacks of a vaporizer are that it is relatively bulky and heavy. In comparison, an injector is compact in size, but it is more expensive than a vaporizer and cannot turn down well, limiting the range of throughput.

CPOx reactors and ATR are designed for low pressure drop and low overall operating pressure because it is expensive to compress air (both in terms of capital cost of the air compressor and operating energy requirements). Typically, these reactors operate within a few tens of kPa of ambient pressure. It is not unusual to have non-uniform flow of the methanol/air/(water) feed stream through the reactor, resulting in hot spots.

3.2.1.1 Energy efficiency

The energy balance in a CPOx reactor is easy to address: the reaction is an exothermic partial oxidation reaction and heat is released in accordance with the reactants and product distribution. Using the same terminology presented above in section 3.1.1.1, the energy efficiency of a CPOx reactor is

$$\eta_{\text{LHVC}} = \frac{REFORM_{\text{th}}}{MeOH_{\text{th}}} \tag{3.12}$$

where η_{LHVC} is the thermal energy efficiency (lower-heating value) of the CPOx reactor, $REFORM_{\text{th}}$ is the lower-heating value of the product reformate stream, and $MeOH_{\text{th}}$ is the lower-heating value of the methanol in the feed stream. Given the simplified and ideal reaction stoichiometry shown in equation 3.10 the following values can be assigned to the variables:

$$REFORM_{\text{th}} = 483.6 \text{ kJ/gmol methanol}$$

$$MeOH_{\text{th}} = 676.1 \text{ kJ/gmol methanol}$$

Thus, $\eta_{\text{LHVC}} = \frac{483.6}{676.1}$ kJ/gmol methanol, or 71% efficiency. In practice, poor selectivity for product hydrogen results in a significantly lower energy efficiency. Also, the above simplified calculation does not account for heat required to vaporize and heat the methanol or for energy required to heat the air fed to the CPOx reactor.

The ATR reaction is more complex due to the fact that neither the steam-reforming reaction nor the CPOx reaction yield, in practice, the simple product distributions discussed above. Considering the more complex and realistic steam-reforming reaction shown as equation 3.8 and the associated energy balance presented in section 3.1.1., a more correct description of the ATR reactor can be put forward.

Combining the steam-reforming stoichiometry (equation 3.8; net enthalpy of reaction is about +168 kJ/mol methanol) and the simplified CPOx stoichiometry (equation 3.10; enthalpy of reaction is about −192 kJ/mol methanol), then solving for zero net enthalpy, yields the following ATR reaction stoichiometry:

$$CH_3OH + 0.71H_2O + 0.14O_2 + 0.57N_2 \rightarrow 0.52\,CO_2 + 0.48\,CO + 2.24H_2 + 0.48\,H_2O + 0.57N_2 \quad (3.13)$$

Following the previous protocol for defining reformer efficiency, the thermal efficiency of an ATR reformer is

$$\eta_{\text{LHVA}} = \frac{REFORM_{\text{th}}}{MeOH_{\text{th}}} \quad (3.14)$$

where η_{LHVC} is the thermal energy efficiency (lower-heating value) of the ATR reformer, $REFORM_{\text{th}}$ is the lower-heating value of the product reformate stream, and $MeOH_{\text{th}}$ is the lower-heating value of the methanol fraction of the feed stream. Based on the reaction stoichiometry shown in equation 3.13,

$$REFORM_{\text{th}} = 677.4 \text{ kJ/gmol methanol}$$

$$MeOH_{\text{th}} = 676.1 \text{ kJ/gmol methanol}$$

Thus, $\eta_{\text{LHVA}} = \frac{677.4}{676.1} =$ kJ/gmol methanol, or 100% efficiency. This is approximately the maximum energy efficiency that can be expected from a methanol ATR since the aforementioned less-than-ideal selectivity of the CPOx reaction will result in a reduction in product hydrogen.

3.2.1.2 CPOx and ATR catalysts

Catalysts for partial oxidation and autothermal reforming of methanol are most commonly based on the platinum-group metals. Rhodium in particular has good activity for catalyzing the oxidation of methanol. Various proprietary commercial formulations are available including from Süd-Chemie, BASF, and Johnson Matthey.

It is customary for CPOx and ATR catalysts to be applied to an engineered support such as a honeycomb monolith or foam. Ceramics (such as cordierite, alumina, and zirconia-toughened alumina) are preferred for the catalyst substrate due to their superior thermal stability at high temperature compared to metals. However, if the reaction temperature is carefully controlled, then stainless steel and Fecralloy (an alloy of iron, chromium, aluminum, and yttrium) may also be used.

3.2.2 *Reactor Designs*

Since CPOx and ATR reactors are heated internally, the reactor design is simpler than for steam reformers. Typically, a cylindrical plug of the catalyst applied to either a monolith or foam ceramic support is inserted into a stainless steel tube shell. The catalyst plug is sealed against the interior shell wall using ceramic felt. Methanol, or methanol/water, feedstock must be vaporized and then mixed with air prior to the leading face of the catalyst plug. Good mixing of all reactants prior to contacting the catalyst is important to achieving uniform reaction and reproducible product distribution.

A cone-shaped plenum prior to the catalyst plug normally suffices for achieving adequate mixing, especially if the reactants are introduced such that turbulent flow results. Initially when the reactor is at or near to room temperature, a glow plug or other means is needed to heat at least a portion of the catalyst leading face to about 200°C to initiate the partial oxidation reaction. Once the CPOx reaction is started, heating the remainder of the reactor is relatively fast since the thermal mass is low.

Depending on the specific reactor design, low space velocity combined with high temperature at the leading face of the catalyst plug can result in flash-back into the mixing plenum. If this occurs, a simple solution is to reduce the volume of the reactor to increase space velocity. Alternatively, a flash arrestor may be placed just ahead of the leading face of the catalyst plug.

It is generally not considered practical to operate either a CPOx reactor or ATR reactor at substantially elevated pressure due to the requirement to compress air. Thus, in practice these reactors are usually operated at <0.5 bar pressure. Any consideration to operation at elevated pressure should be approached cautiously since air oxidation reactions may proceed with unpredictable results—perhaps resulting in severely overheating the reactor.

3.3 Oxidative Methanol Reforming

In an effort to overcome the nitrogen dilution associated with using air as the oxidant in methanol reforming (ATR and CPOx), hydrogen peroxide has been investigated as the oxidant. The reaction is catalyzed by transition metals. Due to the extreme oxidative reactivity of pure (100%) hydrogen peroxide, the reaction is carried out with the addition of water. Still, hydrogen peroxide of concentration from 40% to 50% is used. The mechanism of reaction is not well understood, but has been proposed to involve the decomposition of hydrogen peroxide to yield active oxygen which quickly partially oxidizes methanol generating heat and liberating some hydrogen. The heat raises the reaction temperature increasing the rate of methanol-steam reforming.

The reactor may use a single catalyst composition, or multiple catalyst compositions.[12] For example, a reactor incorporating a catalyst at the inlet that causes hydrogen peroxide decomposition and partial oxidation of methanol followed by a reforming catalyst has been described in a US Patent.[13] The hydrogen peroxide should not contain stabilizers as they may cause an inhibition in initial decomposition of hydrogen peroxide.

Since air is not required, the balance-of-plant is simpler than a conventional ATR. However, it is not advisable to premix methanol and hydrogen peroxide so two liquid delivery systems are required. If the ratio of methanol to hydrogen peroxide is fixed over the entire turn-up/turn-down range, then a two-head liquid pump may be used with one head for methanol delivery and the other head for water delivery. The ratio of pump-head displacement will dictate the molar ratio of methanol to hydrogen peroxide.

There is no need to add water to the methanol as a surplus of water is delivered along with the hydrogen peroxide. A severe drawback to this system is that the reaction of concentrated hydrogen peroxide with methanol in the presence of a catalyst is vigorous and can be unpredictable, leading to runaway thermal excursions and explosions. Also, storage of concentrated (>40%) hydrogen peroxide is usually regulated which may make it difficult for customers to get permission to install hydrogen peroxide storage tanks in some areas. For these reasons, this interesting methanol reforming chemistry has not garnered much attention.

3.4 Commercial Sources and Technical Challenges

Despite the extensive knowledge base for methanol reforming, and its recognized utility in fuel cell systems, there are very few commercial sources of reformers and reformer subsystems. It appears that most commercial development of methanol reforming reactors (with and without hydrogen purification) has been carried out by fuel cell companies that have chosen to keep the technology captive rather than trying to capitalize on the investment by selling it to other companies, possibly competitors.

Four companies that are offering methanol-steam reformers coupled with membrane-based hydrogen purification (so-called fuel processors because reforming is packaged with hydrogen purification) are Element One or E1, (Bend, Oregon, USA), Genesis Fueltech (Spokane, Washington, USA), Hy9 Corp. (Hopkinton, MA, USA), and REB Research and Consulting (Detroit, Michigan, USA). Green Hydrotec (Tao-Yuan, Taiwan) also offers commercial methanol-steam reformers without additional hydrogen purification, which may be more interesting for system integrators. In addition, based on product specification sheets published[14] by Green Hydrotec, it appears they can provide supplemental WGS reactors and palladium-membrane purifiers to achieve a higher degree of hydrogen purity suitable for either high-temperature PEMFC systems or low-temperature PEMFC systems.

Fuel-cell-system integrators focused on product commercialization still have challenges related to the methanol-steam reforming subsystem. These challenges include the following:

- *Cost reduction*. Like all components and subsystems in the fully integrated fuel cell product, the cost of the methanol reformers must be driven down to an acceptable value. Success may be achieved by using more traditional and basic technological approaches rather than the most elegant solutions.

- *Rapid cold start-up*. Many (but not all) applications favor a very rapid start from a cold or ambient condition. The thermal mass of the reformer and associated purification module places limits on how fast the reformer subsystem can be heated to its operating temperature; start-up times in the range of 15 minutes to 30 minutes are normal. Reducing mass and using hybrid solutions (e.g., using an internal battery or supper capacitor to provide electrical power to the external load during start-up) will need to be achieved to be successful.
- *Improved durability*. Depending on the target application, the fuel cell system and methanol reformer subsystem may be required to meet acceptable performance specifications after an operational lifetime of 1,000 hours to >10,000 hours. Reforming catalyst durability is only one aspect of this challenge; it is equally important to ensure that the liquid-feed pump, burner, igniter, and other components have adequate durability and that flow channels and orifices do not plug.
- *Fuel composition*. The preferred blend of methanol and water is dependent on whether the reformer is to produce reformate for a high-temperature PEMFC system or pure hydrogen for a low-temperature PEMFC system. The former favors a mix that has a greater proportion of water (e.g., 40 wt% to 45 wt% water) to drive CO conversion; whereas the later favors close to a 1:1 molar ratio of methanol and water (e.g., 37 wt% to 38 wt% water).
- *Energy efficiency.* The selectivity of the reforming reactions should be optimized for the selected chemical process in order to maximize energy efficiency. There is opportunity for improved catalysts that can deliver high selectivity to the desired product (H_2), fast kinetics, and good durability.

As the above challenges are resolved, we can expect to see greater commercial product success for methanol fuel reformers.

Notes

1. J.C. Brown and E. Gulari, "Hydrogen production from methanol decomposition over Pt/Al_2O_3 and ceria promoted Pt/Al_2O_3 catalysts," *Catal. Commun.*, 5 (2004) pp. 431–436
2. O. Ilinich, Y. Liu, C. Castellano, G. Koermer, A. Moinin, and R. Farrauto, "A new palladium-based catalyst for methanol steam reforming in a miniature fuel cell power source," *Platinum Metals Rev.*, 52, No. 3 (2008) pp. 134–143
3. D.R. Palo, R.A. Dagle, and J.D. Holladay, "Methanol steam reforming for hydrogen production," *Chem. Rev.*, 107 (2007) pp. 3992–4021
4. K. Shah, and R.S. Besser, "Understanding thermal integration issues and heat loss pathways in a planar microscale fuel processor: demonstration of an integrated silicon microreactor-based methanol steam reformer," *Chem. Eng. J.*, 135S (2008) pp. S46–S56
5. J. Bravo, A. Karim, T. Conant, G. Lopez, and A. Datye, "Wall coating of a CuO/ZnO/ Al_2O_3 methanol steam reforming catalyst for microchannel reformers," *Chem. Eng. J.* 101 (2004) pp. 113–121
6. A. Kundu, J.M. Park, J.E. Ahn, S.S. Park, Y.G. Shul, and H.S. Han, "Microchannel reactor for steam reforming of methanol," *Fuel*, 86 (2007) pp. 1331–1336
7. J.M. Sohn, Y.C. Byun, J.Y. Cho, J. Choe, and K.H. Song, "Development of the integrated methanol fuel processor using microchannel patterned devices and its performance for steam reforming of methanol," *Int. J. Hydrogen Energy*, 32 (2007) pp. 5103–5108
8. A. Kundu, J.E. Ahn, S.-S. Park, Y.G. Shul, and H.S. Han, "Process intensification by microchannel reactor for steam reforming of methanol," *Chem. Eng. J.*, 135 (2008) pp. 113–119
9. M.J. Castaldi, R.S. Boorse, S. Roychoudhury, and P.V. Menacherry, "A compact, lightweight, fast-response preferential oxidation reactor for PEM fuel cell applications," http://www.precision-combustion.com/proxpaper.pdf, accessed on September 28, 2009
10. S. Patel and K.K. Pant, "Selective production of hydrogen via oxidative steam reforming of methanol using Cu-Zn-Ce-Al oxide catalysts," *Chem. Eng. Sci.*, 62 (2007) pp. 5436–5443
11. K.-S. Choi, H.-M. Kim, J.L. Dorr, H.C. Yoon, and P.A. Erickson, "Equilibrium model validation through the experiments of methanol autothermal reformation," *Int. J. Hydrogen Energy*, 33 (2008) pp. 7039–7047
12. Y. Kawamura, T. Ishida, W. Tezuka, and A. Igarashi, "Hydrogen production by oxidative methanol reforming with various oxidants over Cu-based catalysts," *Chem. Eng. Sci.*, 63 (2008) pp. 5042–5047
13. A.R. Oroskar, K.M. Vanden Bussche, L.T. Nemeth, and C.M. Rayner, *Process for the production of hydrogen*, US Patent #7,022,306 (April 4, 2006)
14. Company specification sheet prepared for the Fuel Cell Expo, Tokyo, Japan (February 2009)

Chapter 4

HYDROGEN PURIFICATION

Although hydrogen is the most abundant element in the universe it is never found in the free, elemental state on Earth. Indeed, the preceding chapters have addressed thermochemical methods for synthesizing hydrogen from methanol (with or without water) for precisely this reason. However, the reaction products invariably include high concentrations of carbon dioxide and carbon monoxide, and perhaps low concentrations, or even trace concentrations, of a wide range of organic byproducts.

Low-temperature PEMFC stacks require a relatively high degree of hydrogen purity. For instance, Ballard specifies[1] that hydrogen for consumption in their low-temperature fuel cell be at least 99.95% pure with <0.2 ppmv carbon monoxide and no more than trace levels (parts-per-billion) of functionalized organic compounds. AFC stacks additionally require very low concentrations of carbon dioxide due to adverse reaction of carbon dioxide with the alkaline electrolyte.

Fuel cells operating at higher temperature (e.g., SOFC and MCFC) are not required to use such a high purity of hydrogen. Even PAFC stacks can operate adequately with hydrogen containing up to a few percent carbon monoxide, but the presence of sulfur compounds and reactive organic contaminants will still cause degradation in fuel cell performance.

Methanol Fuel Cell Systems: Advancing Towards Commercialization
By Dave Edlund

www.panstanford.com

4.1 Hydrogen Purification Applied to Methanol Reformers

One of the reasons why methanol is a preferred fuel for reforming to hydrogen is because there are few side reactions and the principle byproducts are carbon monoxide, carbon dioxide, and methane. Trace concentrations of formaldehyde and formic acid can be formed, but this is rare and usually only occurs under off-design conditions (e.g., very low reforming temperature). And since methanol is synthesized using conventional C1 chemical reactions and catalysts, sulfur is absent from commercial methanol (unlike petroleum-based liquid fuels, which always are contaminated with organosulfur compounds).

There are two fundamentally different approaches that are used for purifying hydrogen from reformed methanol: purification by chemical methods and purification by membrane separation. Because reformed methanol fuel cell systems tend to be less than 10 kW_e in scale, pressure swing adsorption (PSA) is not economically viable and will not be discussed in detail.[2]

4.2 Chemical Purification Methods

Since carbon monoxide is the most significant impurity in methanol-derived reformate, chemical methods to remove (or greatly reduce the concentration of) carbon monoxide can be employed with generally positive results. These methods include the following:

- Preferential oxidation of carbon monoxide to carbon dioxide (PrOX)*

$$2CO + O_2 \rightarrow 2CO_2 \tag{4.1}$$

- Selective methanation of carbon monoxide to methane (SelMETH)

$$CO + 3H_2 \rightarrow CH_4 + H_2O \tag{4.2}$$

- Water-gas shift reaction (WGS)

$$CO + H_2O \rightarrow CO_2 + H_2 \tag{4.3}$$

---

* This process may also be correctly referred to as selective oxidation or SelOX.

However, it must always be remembered that these chemical methods target only the carbon monoxide content of the reformate stream. If other deleterious organic contaminants are present (e.g., formic acid, formaldehyde, or unreacted methanol) they will not be removed from the product stream. Furthermore, such contaminants could conceivably poison the catalysts that are used to accelerate the reaction kinetics of the chemical purification process.

Like all chemical reactions, the above purification processes are governed by thermodynamic equilibrium and chemical kinetics. Generally, the acceptable range of operating temperature within the reactor must be controlled tightly to avoid unwanted side reactions (e.g., oxidation of hydrogen in the case of PrOX, or methanation of carbon dioxide in the case of SelMETH) which will occur to a significant degree if the reaction temperature is too high. On the other hand, if the reaction temperature is too low, the reaction rate (chemical kinetics) becomes too slow to achieve the minimum degree of carbon monoxide conversion, and the product stream fails to meet minimum purity requirements with respect to carbon monoxide.

Another drawback to chemical purification methods is that the process, as implemented in a reformer subsystem, does not follow variable reactor output very well. This is because the purification reactor size is fixed, and thus the amount of catalyst is fixed as well. The entire product stream flows through the purification reactor. In the case of minimum reformate output from the methanol reformer, the flow rate of reformate gas into the purification process (e.g., through the PrOX, SelMETH, or WGS reactor) is at the minimum value and the residence time within the purification reactor is at its maximum value. Maximizing residence time can lead to loss of selectivity in the process.

When the methanol reformer is at maximum output, the flow rate of reformate gas into the purification process is at its maximum value and the residence time within the purification reactor is at its minimum value. This condition can lead to failure to remove sufficient carbon monoxide from the product stream (reduced carbon monoxide conversion).

Compounding the operation of these chemical reactors is the heat load on the purification reactor: both PrOX and SelMETH are highly exothermic processes, whereas the WGS is slightly exothermic. So the heat load, or the amount of thermal energy that must be dissipated from the purification reactor, also varies with the output of the methanol reformer (both chemical composition and flow rate).

All of these factors combine to make chemical purification processes difficult to control in a dynamic operating mode. If the methanol reformer is operating at a fixed output rate or within a narrow turn-down range, then the success of implementing PrOX, SelMETH, and WGS as carbon monoxide removal processes is improved.

4.2.1 *Water-Gas Shift Reaction*

The WGS reaction has long been used as a means for reducing the carbon monoxide content of a gas stream and simultaneously increasing the hydrogen concentration. The reaction (equation 4.3) has been extensively studied and it is widely employed in the manufacture of basic chemicals.[3] The reaction is mildly exothermic (ΔH = −41.1 kJ/mol carbon monoxide) and to achieve high conversion to carbon dioxide the reaction temperature should be low; preferably below 250°C. Even so, to obtain carbon monoxide concentrations in reformate that are less than 3% may require addition of excess water rather than using a feed stream containing a 1:1 molar ratio of methanol to water.

In the manufacture of chemicals, a high-temperature (HT) shift reactor is commonly used in conjunction with a low-temperature (LT) shift reactor. The HT shift reactor, placed upstream of the LT shift reactor, yields rapid kinetics for the WGS reaction, but less than the desired degree of carbon monoxide conversion. The LT shift reactor is capable of driving carbon monoxide conversion to the target value, but the kinetics in the LT shift reactor are slow, equating to a large reactor. Ultimately it is the slow kinetics of the WGS reaction as reaction temperature is reduced that limits the usefulness of the WGS reaction in achieving very low concentrations of carbon monoxide; slow reaction kinetics implies large and expensive reactors.

The WGS equilibrium expression is

$$K_{\text{eq}} = \frac{(P_{H_2}P_{CO_2})}{(P_{H_2O}P_{CO})} \tag{4.4}$$

where K_{eq} is the equilibrium constant calculated from the partial pressures of the reactants (carbon monoxide and water) and the products (hydrogen and carbon dioxide). The equilibrium constant has been tabulated as a function of temperature from 200°C to 1195°C in the *Catalyst Handbook*.[4]

Due to the WGS equilibrium limitations imposed by the reaction temperature as well as the relative concentration of hydrogen, carbon dioxide, and water, it is not practical to drive carbon monoxide conversion to the low concentrations necessary for PEMFC applications (≤10 ppmv and, preferably, ≤1 ppmv). However, in the case of PAFC and high-temperature PEMFC systems, where the fuel cell stack operates at 140°C to 180°C, the carbon monoxide concentration in the anode hydrogen fuel stream may be as high as 3 vol%. In these specific cases, the WGS reaction may be effective at achieving a carbon monoxide concentration in the product hydrogen-rich reformate stream that is ≤3 vol%. It is especially helpful if excess water is present, such that the methanol/water mix might contain as little as 50–55 wt% methanol (64 wt% methanol in water represents a 1:1 molar blend of methanol and water).

Adding excess water to the reaction comes at a cost of additional energy input into the reaction system to vaporize and heat the excess water. Thermal energy can be recovered from the exit stream (reformate product stream), especially if a condensing heat exchanger is employed to recover the heat of vaporization of the water. However, heat exchangers are expensive and require a suitable low-temperature fluid on the cold side of the heat exchanger.

Because of the industrial utility of the WGS reaction, catalysts are readily available from many manufacturers (e.g., BASF, Johnson Matthey, Süd-Chemie, and Tanaka Metals) in a variety of shapes, sizes, and compositions. Copper-zinc formulations have high activity as LT shift catalysts, but suffer from poor stability combined with being pyrophoric. Precious metal formulations are available, and they are stable in the active reduced state even when exposed to air (i.e., they are not pyrophoric), but these formulations are inherently more expensive than are base-metal formulations. Recently, BASF and researchers at Pacific Northwest national Laboratories (Richland, WA) have developed different Pd-Zn catalyst formulations differing in proprietary promoters. The Pd-Zn catalysts are said to offer activity equal to or exceeding traditional Cu-Zn compositions, yet better stability.[5,6]

Because of the challenge in achieving ppm levels of carbon monoxide from the WGS reactor, this purification approach is generally not applied to low-temperature PEMFC applications. However, it does have merit for high-temperature PEMFC applications. One company using this purification approach is WS Reformer GmbH (Renningen, Germany). Although their FLOX reactor is designed for natural gas and propane/butane, in principle the same approach could be applied to methanol reforming.

4.2.2 *Preferential Oxidation*

The WGS reaction is effective in driving carbon monoxide to low concentrations if an excess of water is present in the feedstream. For example, a methanol/water mixture containing 45 wt% methanol is a molar ratio of 2.17 moles water to 1 mole of methanol. Assuming 95% methanol conversion in the reforming reactor and further assuming that the reformate stream is at equilibrium following the WGS reactor (exit temperature of 200°C), the product wet reformate stream will contain approximately 0.2 vol% carbon monoxide (2,000 ppmv carbon monoxide) plus about 56 vol% hydrogen, about 19 vol% carbon dioxide, and about 24 vol% water. This concentration of carbon monoxide is far too high for use in low-temperature PEM fuel cells.

Of all the chemical purification methods that have been considered for use with methanol reformers, PrOX has been the most successful.[7,8,9] Since the reaction is strongly exothermic ($\Delta H = -282.99$ kJ/mol carbon monoxide) PrOX is usually employed downstream of a WGS reactor. Table 4.1 shows the temperature rise expected as a result of selective oxidation of carbon monoxide as a function of the carbon monoxide concentration. The assumption is that the

Table 4.1 Calculated temperature rise for a reformate stream of variable carbon monoxide concentration subjected to PrOX (reformate stream contains approximately 56 vol% hydrogen)*

Carbon Monoxide Concentration (ppmv)	Calculated Temperature Rise (°C)	Percent Hydrogen Oxidized	Calculated Temperature Rise (°C)
200	1.71	0.1	7.33
500	4.29	1.0	73.30
1,000	8.58	5.0	366.00
2,000	17.20		
10,000	85.80		

*Added temperature increase due to combustion of hydrogen shown in the columns on the right.

reformate stream has an average heat capacity of 33 J/K•mol, and that there is no oxidation of hydrogen, an undesirable side reaction. Assuming the PrOX reaction is 100% selective for carbon monoxide oxidation (no oxidation of hydrogen) the temperature rise for a reformate stream containing 2,000 ppmv carbon monoxide is almost 20°C; substantial given that it is important to control the operating temperature of the PrOX reactor within a fairly narrow band of about 30°C to 40°C for optimal performance.

The temperature rise is manageable at low concentrations of carbon monoxide (<1,000 ppmv), but such low levels are difficult to achieve in a WGS reactor except when a large excess of water is present. At concentrations of carbon monoxide greater than 2,000 ppmv the temperature rise is significant, making it challenging to dissipate the heat. If the heat cannot be completely dissipated, then there will be a loss in selectivity; meaning that hydrogen will also be oxidized contributing further to excessive temperature rise. The result can very quickly become a runaway reaction.

The table also shows the temperature rise that is calculated for oxidation of a small portion of hydrogen. This is unavoidable, but must be managed with suitable catalyst selection, temperature control, and contact time in the PrOX reactor (space velocity). PrOX reactors usually operate at very high space velocity (up to 440,000

hr^{-1} has been reported[10]) to manage byproduct hydrogen oxidation. Catalysts that have been reported to be effective include platinum, gold, ruthenium, and rhodium.

For these reasons, PrOX reactors tend to be complex to control. Although PrOX has been the subject of many development efforts and there have been several high-profile demonstration projects using PrOX reactors and methanol reformers (for instance, Daimler Benz used a methanol reformer followed by a WGS reactor and PrOX in many of their prototype fuel-cell cars), this purification process has not gained commercial traction.

4.2.3 *Selective Methanation*

Conceptually, SelMETH offers system simplification compared to PrOX since the reactants (carbon monoxide and hydrogen) are already present in the feed stream (unlike PrOX in which air must be added to the feed stream). The reaction is exothermic ($\Delta H = -206.21$ kJ/mol carbon monoxide) and the calculated temperature rise as a function of carbon monoxide concentration in the feed stream is given in Table 4.2 (assuming the reformate stream has an average heat capacity of 33 J/K•mol, and there is no methanation of carbon dioxide). As with the above discussion of PrOX, it is assumed

Table 4.2 Calculated temperature rise for a reformate stream of variable carbon monoxide concentration subjected to SelMETH (reformate stream contains approximately 19 vol% carbon dioxide)*

Carbon Monoxide Concentration (ppmv)	Calculated Temperature Rise (°C)	Percent Carbon Dioxide Methanated	Calculated Temperature Rise (°C)
200	1.25	0.1	5.0
500	3.12	1.0	50.0
1,000	6.25	5.0	250.0
2,000	12.50		
10,000	62.50		

*Added temperature increase due to methanation of carbon dioxide shown in the columns on the right.

that 95% methanol conversion is achieved in the reforming reactor and further assummed that the reformate stream is at equilibrium following the WGS reactor (exit temperature of 200°C), the product wet reformate stream will contain approximately 0.2 vol% carbon monoxide (2,000 ppmv carbon monoxide) plus about 56 vol% hydrogen, about 19 vol% carbon dioxide, and about 24 vol% water.

The expected temperature rise is modest up to 1,000 ppmv carbon monoxide; but is significant at higher concentrations of carbon monoxide. It is important to maintain thermal control of the SelMETH reactor since selectivity is sacrificed at higher temperatures. The methanation of carbon dioxide is also exothermic, and the calculated temperature rise upon methanation of carbon dioxide is also presented in Table 4.2 (right side). Methanation of carbon dioxide is undesirable because three moles of hydrogen are consumed for every mole of carbon dioxide that is converted to methane, and methanation of carbon dioxide contributes significantly to excessive increase in temperature.

SelMETH reactors will typically operate in a narrow range of temperature, for instance 175°C to 250°C. Ruthenium dispersed as nano particles is a preferred catalyst.[11,12,13] Since operating temperatures are low (implying slow kinetics), space velocity is also relatively low, in the range of 500 to 3,000 hr^{-1}. SelMETH has been the subject of development interest since at least the mid-1960s and yet has failed to see adoption for commercial applications.

4.2.4 *Membrane Purification Methods*

Unlike chemical methods of purifying hydrogen, membrane-based separation processes (and PSA as well) will remove essentially all impurities from the feed stream resulting in a substantial increase in hydrogen purity. In this sense, membrane purification can be considered a universal purification process for hydrogen. A generic membrane process for purifying hydrogen is shown in Fig. 4.1.

Membrane purification processes may utilize polymeric membranes, metallic membranes, or ceramic membranes; and the membranes may be either dense or nano-porous. But in all cases

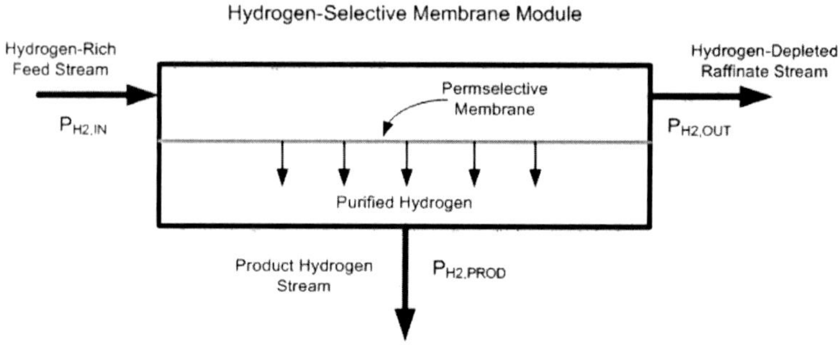

Figure 4.1 Typical membrane process for purifying hydrogen. The membrane may be any hydrogen-selective material. P_{H_2} refers to hydrogen partial pressure and the subscripts IN, OUT, and PROD refers to the incoming feed, the outgoing raffinate, and the product streams, respectively.

selective permeation of hydrogen† across the membrane is desired such that impurities are left on one side of the membrane and purified hydrogen on the opposite side of the membrane. A net transport of hydrogen from one side of the permselective membrane to the other side occurs due to a concentration gradient of hydrogen across the membrane. This concentration gradient is most often the direct result of a pressure differential across the membrane: the hydrogen concentration within the membrane is higher at the surface exposed to higher partial pressure of hydrogen.

The volumetric flow rate at which a gas (e.g., hydrogen) permeates a membrane is called the flux (J_i) and it is related to membrane thickness (l), permeability (P_i), and differential pressure (ΔP), as shown by equation 4.5.

$$J_i = P_i\,(\Delta P/l) \tag{4.5}$$

where the subscript i refers to the specific gas that permeates the membrane and ΔP is the partial pressure of gas i in the feed stream minus the partial pressure of gas i in the permeate stream (for pure-gas permeation; ΔP has a more complex description for mixed-gas

†It is possible to design a purifier such that one or more non-hydrogen species selectively permeate the membrane, leaving a hydrogen-enriched product stream at high pressure. However, this approach has not been shown to yield hydrogen of sufficient purity for PEMFC systems.

permeation—see section 4.2.4.2). The common unit for the permeability constant P_i is the Barrer: 10^{-10} cm^3 (STP) • cm / (cm^2 • s • cmHg).

Hydrogen permeates dense (i.e., non-porous) membranes by a solution-diffusion mechanism. In the case of porous membranes, hydrogen may permeate by gas-phase Knudsen diffusion or by a surface transport mechanism in which hydrogen weakly adsorbs to the pore surface and migrates through the pore structure of the membrane. Since porous membranes cannot yield the degree of hydrogen purity that is required for low-temperature fuel cells, our discussion on membrane purification will focus on dense polymeric and metallic membranes.

Considering solution-diffusion permeation mechanisms, another useful relationship is

$$P_i = D_i \bullet S_i \tag{4.6}$$

where D_i is the diffusivity of the permeating gas in the bulk membrane and S_i is the solubility of the permeating gas in the bulk membrane. Clearly, either increasing the diffusivity term or the solubility term results in an increase in permeability across the membrane. The diffusivity term D_i is a consequence of the molecular interactions of hydrogen with the membrane matrix (chemical composition and molecular or atomic structure), and it is not significantly influenced by the hydrogen partial pressure at the membrane surface. On the other hand, the solubility of hydrogen in the membrane is directly influenced by the hydrogen partial pressure such that a higher pressure of hydrogen results in increased hydrogen solubility in the membrane.

Regardless of the type of membrane, hydrogen recovery is defined as

$$\text{Recovery} = (M_{H_2,p}/t)/(M_{H_2,f}/t) \tag{4.7}$$

where $M_{H_2,p}$ and $M_{H_2,f}$ are the moles of hydrogen in the permeate stream and in the feed stream, respectively, and t is a constant unit of time (e.g., one minute or one hour). Usually, high hydrogen recovery is desired to reduce the operating costs (cost of feedstock to make a given amount of hydrogen). However, as the hydrogen recovery increases, the driving force for permeation of hydrogen across the membrane diminishes. Thus, there is a practical limit to the degree of hydrogen recovery as approaching the theoretical limit of hydrogen recovery results in an exponential increase in required membrane area and associated capital cost.

Provided that hydrogen partial pressure is the driving force for selective permeation across the membrane, then under no circumstances can the pressure of the permeate hydrogen (product stream) exceed the minimum partial-pressure of hydrogen in the feed stream passing over the membrane. Considering the simplified membrane module schematic shown in Fig. 4.1, the minimum feed-stream hydrogen partial pressure occurs where the feed exits the module ($P_{H_2,OUT}$) and is equivalent to the hydrogen partial pressure in the raffinate stream.

4.2.4.1 Palladium-alloy membranes

Due to the low hydrogen selectivity of polymeric membranes and porous membranes, these membranes have not been commercially successful for purifying hydrogen, produced by reforming alcohol and hydrocarbon feedstocks, for fuel cell applications. However, palladium-alloy membranes offer infinite selectivity for hydrogen, and these alloys have been enjoying commercial success since the early 1960s. Additionally, palladium-alloy membranes are operated at elevated temperature; most commonly in the range of 300°C to 400°C, and direct integration with a methanol reformer is not only possible, but preferred.

There are several references the reader should consult for a detailed study of dense-metal membranes and module design.[14–16] However, this section will address some of the more important issues.

Because palladium-alloy membranes (in general terms, dense-metal membranes that are permeable to hydrogen since a great many metal alloys are permeable to hydrogen) are infinitely selective to hydrogen, the purity of the permeate hydrogen does not vary with the trans-membrane pressure driving force. In other words, even if the partial pressure of non-hydrogen impurities is high relative to the hydrogen partial pressure, the purity of the hydrogen that permeates the membrane will not be compromised provided the membrane does not suffer pin holes, tears or other defects that allows the passage of impurities.

The reason that only hydrogen permeates these dense metal membranes is because hydrogen—a chemical element with metallic properties—actually alloys with the membrane alloy.[17,18] In doing so, the molecular hydrogen dissociates to atomic hydrogen, and it is atomic hydrogen that dissolves in the metal alloy and diffuses across the membrane. This is a chemical reaction that is unique to hydrogen among all gasses. Palladium is remarkable in that no other element has such a high affinity for hydrogen. However, pure palladium is a poor choice for hydrogen purification membranes because palladium reacts with hydrogen to form a metallic hydride. When this occurs, the crystal structure of the metal abruptly changes and the result is extreme physical distortion of the palladium membrane, often causing tears and ruptures. The solution is to alloy palladium with other metallic elements such that stable metal hydrides do not form.[19] The most familiar commercial alloys for hydrogen purification membranes are Pd-40Cu and Pd-23Ag.

There are two basic types of hydrogen purification membrane modules that utilize palladium-alloy membranes: planar and tubular. Planar, or foil, membranes are easily fabricated by conventional metal-foil rolling mills. Foil can be rolled very thin without introducing defects so long as the initial alloy ingot is high quality. It is common to purchase Pd-40Cu foil that is up to 20 cm wide and 15 micron thick, and foil as thin as 5 microns can be produced commercially. Tubular membranes made by a drawing process cannot be made with walls this thin; typically tubular membranes will have a wall thickness that is between 50 and 100 microns thick.

There has been considerable development work aimed at making tubular membranes with a very thin permselective palladium alloy layer by chemical deposition of palladium with either gold, silver, or copper onto a porous tube (either metal or ceramic).[20] However, these efforts have not yet yielded commercial products.

Flat foil membranes are built into modules using picture-frame construction methods common to any planar membrane or filtration medium. Frames define the feed channels and permeate collection spaces. Since the foil membranes are very thin, it is imperative that a robust support structure be used to support the membrane against the transmembrane pressure, which may easily be as high as 200 psig. Porous metal and porous ceramic inserts, and woven mesh are most commonly used as support structures.

Seals may be made using flexible graphite gaskets, solid-state diffusion bonding, brazing, and welding. However, gasketing is by far the least expensive sealing method.

In contrast, tubular membranes are invariably brazed to a header. They are constructed analogously to a shell-and-tube heat exchanger using as few as one coiled tube or many tubes. The diameter of the membrane tubes is usually about 3 to 8 mm. Although the feed stream may flow either through the bore of the tubular membranes or over the outer tube surface, much better performance is achieved if the feed passes down the bore of the membrane. This is because mass transfer is far better to the inner wall; achieving efficient mass transfer through the shell side (the outer surface of the tubes) is difficult and expensive. An example of a commercial tubular membrane module is offered by Power+Energy (USA) and is shown in Fig. 4.2.

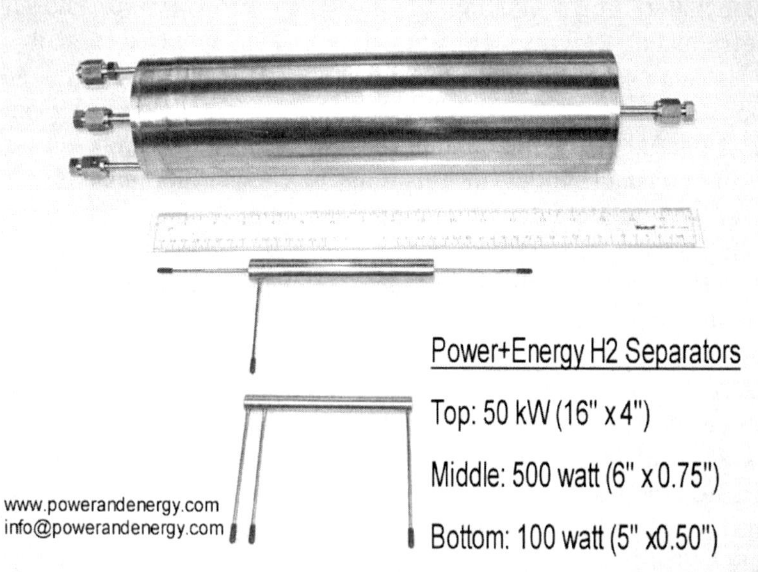

Figure 4.2 Sample tubular palladium-alloy membrane modules sold by Power+Energy (photo courtesy of Power+Energy). See also Color Insert.

Tubular membrane modules are bulkier than planar membrane modules with a similar purification capacity. However, planar membrane modules are usually considerably heavier than tubular modules due to the mechanical requirement for relatively thick steel endplates with planar membrane modules. A planar membrane module would be expected to weigh two or even three times the weight of a module using tubular membranes (both with equivalent hydrogen purification capacity).

4.2.4.2 Modeling palladium-alloy membrane performance

The general equation for gas permeation through a membrane discussed above (equation 4.5) must be modified to accurately describe hydrogen permeation through a dense metallic membrane. The chemical reactions that occur resulting in permeation of hydrogen through any dense metal following a solution-diffusion mechanism are

$$H_{2(g)} + M_{(s)} \leftrightarrow 2\ H\text{-}M_{(s)} \qquad (4.8)$$

$$2H\text{-}M_{(s)} \leftrightarrow 2\ (H_d)M \qquad (4.9)$$

where $H_{2(g)}$ is gaseous molecular hydrogen; $M_{(s)}$ is the metal membrane surface, $H\text{-}M_{(s)}$ represents a hydrogen atom chemisorbed onto the metal membrane surface, and $(H_d)M$ is the hydrogen atom dissolved within the metal membrane matrix. All of these steps are reversible and facile at operating temperatures >200°C.

Since each molecule of gas-phase molecular hydrogen dissociates to yield two hydrogen atoms dissolved in the metal membrane matrix, the equation for flux dependence on pressure is modified as follows:

$$J_i = P_i(\Delta(P^{0.5})/l) \qquad (4.10)$$

where the variables have the same meaning as presented for equation 4.5. In other words, for a pure hydrogen system the rate of permeation across the membrane is proportional to the difference between the square-root of the feed hydrogen pressure and the square-root of the permeate hydrogen pressure. This is assuming ideal behavior.

In practice, there are many variables that cause the pressure dependence to vary somewhat. These variables include the composition of the membrane surface, the degree of roughness of the membrane surface, presence of trace contaminants that may coat the membrane surface, operating temperature, and operating pressure. Anything that affects the rate of hydrogen dissociating and dissolving within the metal membrane (equations 4.8 and 4.9) will have the effect of causing the value of the pressure exponent to increase from 0.5 to approach 1. Thus, a more general expression is

$$J_i = P_i(\Delta(P^n)/l) \tag{4.11}$$

where the exponent n is a value between 0.5 and 1.0 and is best determined experimentally. Once measured, the value of the exponent n provides insight into the rate-limiting mechanism for hydrogen permeation. A value for n that is close to 1.0 implies that the diffusion of hydrogen across the membrane is fast and the surface reactions are slow, whereas a value for n that is close to 0.5 implies that the diffusion across the membrane is slow and the surface reactions are fast. The former case is expected for very thin metal membranes and the latter case is expected for thick metal membranes.

The foregoing equations relating flux to pressure differential have assumed that the feed stream is pure hydrogen; not a practical application. Borrowing theory from heat transfer, the rate of hydrogen permeation that occurs in a membrane module when hydrogen is mixed with other gases is described well by the ln-mean driving force. This expression takes the form shown below:

$$F = \frac{[(P_{H_2,IN})^n - (P_{H_2,PROD})^n] - [(P_{H_2,OUT})^n - (P_{H_2,PROD})^n]}{\ln[(P_{H_2,IN})^n - (P_{H_2,PROD})^n / (P_{H_2,OUT})^n - (P_{H_2,PROD})^n]} \tag{4.12}$$

where F is the ln-mean pressure driving force and the other variables have the meaning defined in Fig. 4.1. The exponent n has the same meaning as is defined above in the context of equation 4.11.

4.2.4.3 Economic considerations

Thinner membranes, if they can be produced free of tears and pin holes, are far more economical than thick membranes since the hydrogen flux increases with decreasing membrane thickness (ref to equation 4.11). A simple analysis of the cost of palladium in a given membrane shows that the membrane cost is roughly proportional to the inverse square of the membrane thickness. Thus a palladium alloy membrane that is ½ the thickness of a reference membrane will yield the same flow of product hydrogen (all other process variables held constant) at ¼ the cost. If the membrane is further reduced to ⅓ the thickness of the reference membrane the cost will be reduced to 1/9 that of the reference membrane. The limitation of this simplified analysis is that it ignores the fabrication cost of making ultra-thin foils.

But considering only the value of palladium in the metal membranes is misleading, since the cost of fabricating palladium alloy membranes—commonly used as foils and thin-walled tubes—may be as much as five times the value of the palladium. Plus there is the cost of the membrane module construction, which naturally scales with the membrane area.

The various parts used in assembling a module using planar membranes can be simply represented as endplates; feed frames (to define the feed flow channel); permeate plates (to define the permeate hydrogen collection region, encloses the membrane support structure); membrane support structure (prevents the membrane from rupturing under high transmembrane differential pressure); gaskets; and bolts (to assemble the module).

The generic parts and their relative cost for a module sized to purify approximately 35 NLm from methanol reformate is shown in Fig. 4.3. The value of palladium incorporated in the membranes represents only 10% of the total cost of the membrane module materials. Fabricating the palladium-alloy foil is the single largest cost, but the combined cost of the module internal parts represents about one third of the total cost.

So when considering the cost impact of additional membranes to achieve a higher hydrogen recovery, the incremental increase in the

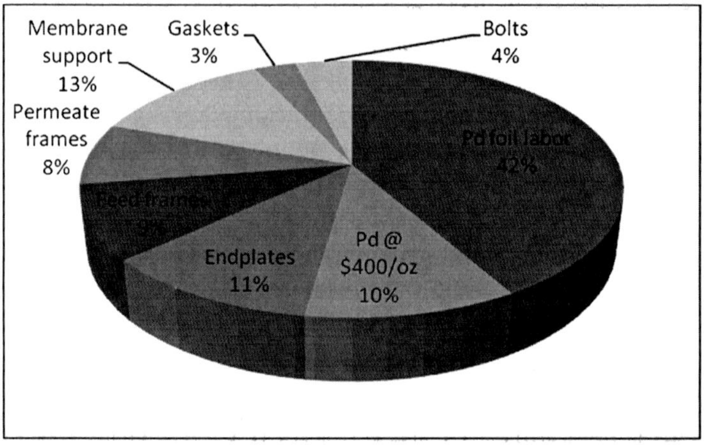

Figure 4.3 Relative cost breakdown (excluding labor for module assembly) for a module using planar Pd-40Cu membranes 15 microns thick and flexible graphite gaskets.

cost of the palladium incorporated in the additional membrane area significantly understates the true additional cost.

Since a membrane module incorporating tubular palladium-alloy membranes would not include feed frames, permeate frames, membrane support structure, endplates, gaskets, and bolts, the cost breakdown for a module using tubular membranes is skewed toward palladium and tube fabrication dominating the total module cost (materials only). This is shown in Fig. 4.4.

However, for an equivalent hydrogen purification capacity, to cost of modules using tubular membrane remains high because the wall thickness of the tube is three to seven times thicker than can be achieved with rolled foil.

Staying with our example of a membrane module sized to deliver 35 NLm hydrogen from methanol reformate, a planar-membrane module should represent material costs of about $2,150 (Pd-40Cu 15 microns thick; very low manufacturing quantity) whereas a module using tubular membranes 50 microns thick (also Pd-40Cu) would be expected to cost about $3,760 (materials only and very low manufacturing volume). Furthermore, the labor content of the planar-membrane module should be less than the labor content of a module using tubular membranes that

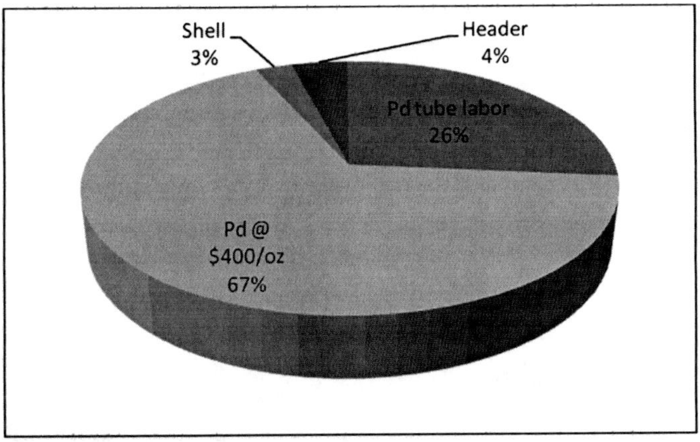

Figure 4.4 Relative cost breakdown (excluding labor for module assembly) for a module using tubular Pd-40Cu membranes 50 microns thick and brazed or welded seals.

must be brazed to a machined header, and then the shell welded closed.

Another factor that influences the cost of a membrane module is the degree of hydrogen recovery that the module is designed to achieve. As hydrogen recovery increases, the driving force for hydrogen permeation across the membrane diminishes. It is unfeasible to realize a hydrogen recovery of 100% because to do so implies extremely low driving force (pressure differential) and a permeate pressure (product hydrogen pressure) that must approach zero. Since the hydrogen partial pressure in the raffinate stream exiting the feed-side of the membrane module cannot be less than the permeate hydrogen pressure, and because the permeate hydrogen pressure must be at a sensible pressure typically greater than ambient pressure, hydrogen recovery is often limited to a value in the range of 80% to 90%.

The important question to evaluate is how much additional membrane area (and cost) is required to approach the limit of hydrogen recovery. An example using a feed stream composition that approximates methanol reformate is depicted in Fig. 4.5 where the calculated membrane area is plotted as a function of hydrogen recovery (the maximum possible hydrogen recovery is 89.45%). The value of the palladium content of the membrane is also plotted.

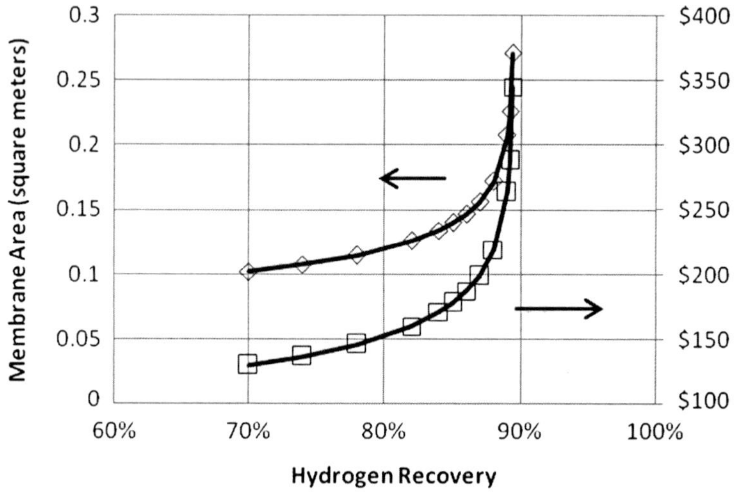

Figure 4.5 Impact of hydrogen recovery on calculated membrane area for the following conditions: feed, 54% hydrogen at 200 psia total pressure; permeate, 22 psia hydrogen; 15 micron thick Pd-40Cu membrane; 35 NLm hydrogen flux; $400/oz. Pd value. Maximum possible hydrogen recovery is 89.45%.

Due to the decreased pressure driving force as hydrogen recovery approaches the theoretical limit, the hydrogen flux decreases causing the amount of required membrane area to exponentially increase.

In this example, which is scaled for a nominal 3 kW PEM fuel cell system, the membrane area, and consequently the palladium value, increases almost three fold on increasing the hydrogen recovery from 70% to 89.4%. As discussed above the module cost will also increase. The important lesson is that increasing hydrogen recovery can be achieved at the expense of additional cost. In other words, operating efficiency can be increased at the expense of higher capital cost.

Other factors to consider when evaluating the optimal hydrogen recovery relate to the intended use of the raffinate stream. The raffinate stream contains all the non-hydrogen impurities that are present in the feed stream plus whatever amount of hydrogen that is not permeated across the membrane. Thus, the raffinate has fuel value, perhaps considerable depending on the amount of hydrogen and carbon monoxide that are present. This stream makes an ideal fuel for firing the methanol reformer, as is discussed in section 4.2.4.5.

4.2.4.4 Membrane durability

Palladium-alloy membranes can be extremely durable. Operating lifetimes as long as 10,000 hours have been reported.[21,22] However, to achieve this goal the membrane must be securely assembled into a module that provides good support to the membrane and utilizes seals that are not overly strained.

If the palladium-alloy membrane is brazed or welded in place, then the brazed or welded joint is a potential point of failure since the composition of the membrane will vary from ideal at the joint. It is natural to think of the composition of the melt pool at a welded joint; but even brazing results in some dissolution of the wetted metal into the molten braze alloy. Normally, the amount of metal that is dissolved is very small and not significant. However, in the case of very thin palladium-alloy membranes, even a small amount of dissolution of the membrane alloy into the liquid phase braze alloy can have a significant impact on the metallurgical composition of the palladium-alloy membrane at the brazed joint.

Problems can occur with joint failure due to differential coefficient of thermal expansion (CTE) at the joint (due to alloy compositional gradation) or failure due to variable hydrogen solubility if hydrogen solubility is high in the palladium alloy (this is the case with palladium-silver alloys but not palladium-copper alloys). As hydrogen dissolves in many alloys, the crystal lattice expands to accommodate dissolved hydrogen atoms at interstitial sites. As other metals from the braze alloy or weld are mixed with the palladium-alloy material, the hydrogen solubility will be effected, and this can cause severe lattice strain just as differential CTE can do so.

Gasketed seals using soft gasketing materials such as flexible graphite are much more forgiving since the gasket can stretch and accommodate a small degree of movement that is not likely to be realized with metallurgical seals. Gasketed seals are an excellent choice for flat (planar) membranes but are difficult to implement with tubular membranes that are often very small in diameter.

Table 4.3 Comparison of the advantages and disadvantages of primary membrane sealing options

Sealing Method	Advantages	Disadvantages
Graphite gasket	• Lowest cost • Tolerates relatively poor fit and surface finish	• Slight leakage inherent • Modules tend to be heavy to apply clamping load on gaskets
Welding	• Hermetic seal	• High cost • Requires good fit • May be difficult to apply to certain shapes
Brazing	• Hermetic seal	• High cost • Requires good fit
Diffusion bonding	• Hermetic seal	• High cost • Requires exceptionally good surface finish and fit

Table 4.3 compares the strengths and weaknesses of the primary membrane sealing methods.

The palladium-alloy membranes must also be either designed to withstand the transmembrane pressure or reinforced with a support structure. Both approaches have been successfully used in commercial products. Tubular membranes are usually made with a thick wall and small internal diameter such that the membrane can withstand a differential pressure of 200 psi. These tubular membranes are almost always fed through the bore with the permeate (low pressure) hydrogen collected from the external surface of the tube.

Planar membranes, on the other hand, are commonly very thin (≤25 microns) and they must be supported with a secondary structure that becomes an integral part of the membrane module. Suitable support structures include porous metal and ceramic plates, and wire mesh sheets. It is preferable that the support structure be either an inert ceramic or a metal that is protected by an inert oxide, nitride, or carbide coating. Whatever material is chosen for the support structure, it must allow hydrogen permeating the membrane to freely flow away from the membrane; otherwise the hydrogen partial pressure at the permeate side of the membrane will be higher than is desirable.

Although aluminum is an attractive material for use in membrane modules due to its low density and relatively high strength (except at elevated temperatures), care must be exercised that palladium-alloy membranes are not in direct contact with aluminum. The metals rapidly inter-diffuse at 400°C and within 10 to 20 hours membrane failure occurs, presumably due to formation of a brittle intermetallic alloy. So using aluminum foam or mesh or perforated sheet stock to support palladium alloy membranes is strongly discouraged. Using aluminum alloys for structural components of membrane modules is acceptable provided that the part is properly designed with consideration to rapid loss in yield strength at operating temperatures >200°C.

As discussed above, the hydrogen flux across the membrane is proportional to the pressure differential across the membrane; i.e., the pressure differential between the partial pressure of hydrogen at the feed-side of the membrane and the partial pressure of hydrogen at the permeate side of the membrane. Experimentally, it is difficult to actually measure the hydrogen partial right at the membrane surface, so usually one measures the total gas pressure and gas composition at some convenient point *near* to the feed surface and permeate surface of the membrane. Extrapolating these measurements to the actual values at the membrane surface is always based on certain assumptions such as uniform gas composition at the feed side of the membrane and low pressure drop at the permeate side of the membrane.

For maximum membrane operational lifetime, conditions that can introduce internal strain in the membrane and or seals should be avoided, or at least minimized. For example, eliminating thermal cycling by maintaining the membrane modules at the operating temperature (usually 300°C to 500°C) will result in almost a complete elimination of strain due to differential CTE. Many times the operating requirements are such that the palladium-alloy membrane module must be repeatedly heated and cooled. If this is the case then materials for fabricating the membrane module should be selected with the goal of minimizing differential CTE. IdaTech addressed this issue successfully and has published a list of preferred steel alloys for use in modules for Pd-40Cu membranes.[23]

Palladium-alloy membranes are known to be poisoned by sulfur compounds, and heavy metals (such as mercury, cadmium, zinc, lead). These contaminants are not normally present in methanol and water, and thus should not be a problem when coupling a palladium-alloy membrane module to a methanol reformer. However, it is important to realize that both the ASME and the ACS specifications for methanol do not regulate sulfur contamination. The IMPCA specification does, however, and is therefore a preferable specification for methanol.

Another impurity that may be encountered is iron oxide—rust. Under operating conditions (hydrogen, 300°C to 500°C) iron oxide is reduced and the resulting metallic iron reacts with palladium-copper membranes causing formation of tiny pin-holes in the membrane. Whether this will also occur with other palladium alloys is uncertain, but it is advisable to avoid contact of thin palladium-alloy membranes with finely divided (nanoscale) metals. Other compounds that should be excluded from contact with palladium-alloy membranes include unsaturated organic compounds (such as alkenes, alkynes, and aromatics) and tars.

The preceding discussion has been concerned with contamination on the membrane surface. Another type of contamination is bulk impurities within the palladium alloy. Normally the specifications for the alloy composition address only the major metal constituents, such as palladium with 40±0.1 wt% copper, or palladium with 23±0.5 wt% silver. The concentration of trace impurities, especially carbon, should also be limited in the alloy specification. Carbon should be at a concentration not greater than 50 ppm.[24]

If IMPCA specification methanol is used for reforming, there should be no problems with contamination. Water used in reforming should be either distilled or de-ionized (meeting or exceeding ASTM Class II specifications). Even with these precautions, there will always be some degree of non-volatile contamination in the methanol and water (so called non-volatile residue or NVR). This material is often predominantly silica and it presents no problem with chemical poisoning or degradation of membrane performance.

NVR can impair performance of either a reformer or purification membrane module if the residue blocks flow passages. For large kilowatt-scale reformers, this is unlikely to occur since flow passages are necessarily large in dimensions (several millimeters or larger). Such is not the case with small—especially portable—methanol reformers and hydrogen purifiers. These are likely to have flow passages than measure a fraction of a millimeter in depth and width. Tiny flow passages of these dimensions are very susceptible to plugging from NVR after as little as 1,000 operating hours.

As much as possible, design of flow passages should be done with consideration to NVR build up over the design lifetime of the apparatus. Alternatively, it is a good practice to place a serviceable filter between the vaporized methanol/water feed stream and the reformer (and hydrogen purifier which is downstream of the reformer). If it is impractical to service the filter (often the case with metal parts that are heated for prolonged periods) then the filter must be adequately sized for the design service life of the reformer.

4.2.4.5 Integration with a methanol reformer

Given the required operating temperature of palladium-alloy membranes (300°C to 500°C) it is natural to examine direct thermal integration of the membrane-based purifier with a methanol reformer. This has been the goal of research efforts since the mid-1990's, perhaps even earlier. In 1993 Ballard Power Systems experimented with a Pd-40Cu membrane module coupled directly to a high-pressure methanol-steam reformer for supplying hydrogen to PEM fuel cells. Since then, Element One, IdaTech, Hy9, REB Research and Consulting, and Genesis Fueltech have all developed commercial methanol reformers that incorporate a palladium-alloy membrane module for delivering pure hydrogen.

Fig. 4.6 shows a generic process diagram for a methanol reformer and thermally integrated membrane-based hydrogen purifier. The dashed line surrounding the burner, vaporizer, reforming reactor, and hydrogen purification membrane represents an insulation enclosure that serves to maintain all the aforementioned items at the desired operating temperature. Combustion exhaust from the burner exits the insulated enclosure, as does the product hydrogen which is cooled via heat exchange with the incoming methanol/water stream.

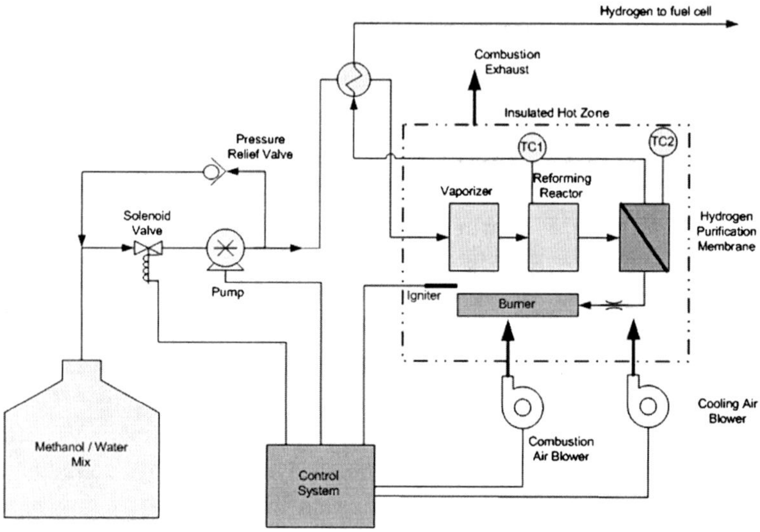

Figure 4.6 P&ID for a methanol-steam reformer coupled directly to a high-temperature membrane module for hydrogen purification.

Combustion air is provided from a suitable blower. Since it is important to maintain the temperature of the reforming reactor and the hydrogen purifier within an acceptable temperature range, cooling air is also supplied via a blower. Increasing the flow of cooling air reduces the temperatures within the insulted box, but at the expense of additional cost and electrical power required to operate the system. An alternative cooling method, employed successfully by Element One, is to turn the combustion blower down (effectively creating a fuel-rich combustion) to lower the temperature of the combustion exhaust and, hence, the temperature of the reformer and the hydrogen purifier. This approach to cooling the reforming reactor results in elimination of the cooling blower.

As shown in Fig. 4.6 the waste gases, or raffinate stream, from the membrane module is directed to the burner. The raffinate stream contains some hydrogen and all the non-hydrogen constituents from the reformate stream, including carbon monoxide. This hot gas stream becomes the fuel that is combusted at the burner resulting in emissions that typically contain <3 ppm NO_x and no particulates.

Inside the insulated enclosure, heat from combustion is consumed by the endothermic processes in the vaporizer and in the reforming reactor. No heat is consumed in the membrane module since the hydrogen separation process is thermally neutral. Ignoring start-up considerations for the moment, it is not necessary to externally heat the membrane module; in fact doing so can easily result in overheating. The membrane module is satisfactorily heated by virtue of the hot reformate gas flowing through the interior of the module, provided it is reasonably well insulated (2.5 to 5 cm of insulation is usually adequate).

The desire to heat the reforming reactor and the membrane module relatively quickly from cold ambient temperature usually means that the combustion exhaust is directed over both the reforming reactor and membrane module. This can make it challenging to manage the membrane module temperature during steady-state operation. How significant this problem is depends to a large degree on the palladium alloy that comprises the hydrogen-selective membranes.

Pd-40Cu is very sensitive to overheating and temperatures greater than 450°C for periods longer than 30 minutes should certainly be avoided. Overheating causes the alloy to undergo a phase transformation from predominately body-centered-cubic (bcc) to predominately face-centered-cubic (fcc). When this transformation occurs, the permeability of the membrane falls substantially; as much as 40%. The process is reversible, but the alloy is very slow to return to the bcc phase.

Other alloys such as the palladium-silver and palladium-gold alloys, as well as Pd-20Cu, continue to improve in permeability as the temperature increases even to 500°C. So regulating the temperature of the membrane module is less of a challenge with these membrane compositions. Still, the structural integrity of the membrane module must be considered and will ultimately define a maximum safe operating temperature that should not be exceeded.

It is tempting to conclude that the hydrogen purification membrane module must be perfect with zero leaks. In fact, some leakage can be easily accommodated by adding a small catalytic methanator in the product hydrogen stream immediately downstream of the membrane module. Either ruthenium or nickel are suitable catalysts and both are available as commercial formulations. Ruthenium has the advantage of being more active, but it is also more expensive than is the nickel catalyst. With a catalytic methanator located at the hydrogen outlet of the membrane module, a pure-gas selectivity‡ as low as 5,000 can be accepted without adversely affecting the product hydrogen purity. This is because the methanator converts carbon monoxide to methane; an inert gas to PEM fuel cells.

Despite the thermal control requirements, the attraction of using highly selective metal membrane for hydrogen purification and thermally coupling this purifier to the methanol reformer yields a simple and affordable design that delivers high-purity hydrogen irrespective of operating state (start-up, full output, partial output) or product hydrogen flow rate.

4.3 Hydrogen Purification for High-Temperature PEMFC

PEM fuel cells that operate at 140°C to 180°C are much more tolerant to carbon monoxide in the anode hydrogen stream than are so-called low-temperature PEM fuel cells. The higher operating temperature provides for tolerance to between 1% and 3% (by volume) carbon monoxide in hydrogen. This is often cited as an advantage of using high-temperature PEM fuel cells over low-temperature PEM fuel cells.

The anode hydrogen must still be essentially free of sulfur compounds and unsaturated or functionalized organic compounds, but this is not a concern when using a methanol reformer. Methanol reformate will typically contain between 3% and 12% carbon monoxide depending on the catalyst used, the operating temperature, and the methanol-to-water ratio. The challenge, then, is to reduce the carbon monoxide content to an acceptable level; high-purity hydrogen is not required.

As previously discussed (see section 3.1.1.2) palladium-zinc methanol-steam reforming catalyst formulations are selective for carbon dioxide formation over carbon monoxide. If the PEM fuel

‡ Pure-gas selectivity is the ratio of product hydrogen flow rate divided by the flow rate of gaseous impurity such as nitrogen or carbon monoxide.

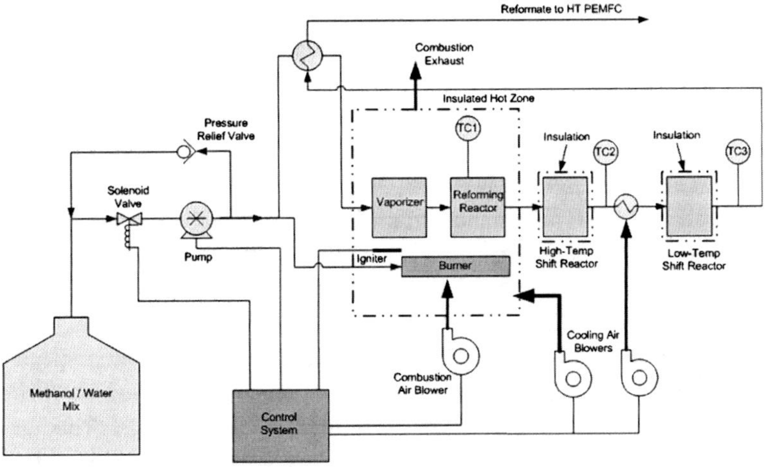

Figure 4.7 P&ID for a methanol-steam reformer using a two-stage WGS reactor to achieve a low concentration of carbon monoxide in the reformate stream.

cell is operated at 180°C, and the methanol reformer temperature is held below 400°C, then it is likely that no further purification of the reformate stream will be necessary. However, other reforming catalysts that have a lower selectivity for carbon dioxide will necessitate partial purification of the product hydrogen to achieve an acceptably low carbon monoxide concentration.

The most practical approach to adjusting the carbon monoxide concentration in the reformate stream is the WGS reaction (discussed above). The reaction is simple with virtually no side reactions, and there are many commercially available catalysts. A typical process schematic is shown in Fig. 4.7.

There are several possible derivations on the schematic above, but in principle it is important to control the output temperature (TC3) since the equilibrium concentration of carbon monoxide is strongly dependent on the temperature within the WGS reactor. Interstage cooling—indicated by the heat exchanger between the HT shift reactor and the LT shift reactor—is necessary to cool the reformate stream. Also notice that unlike the P&ID shown in Fig. 4.6, fuel to the burner is provided directly from the methanol/water feed stream.

4.3.1 *Economic Considerations*

It is tempting to conclude that since a less rigorous purification of hydrogen is acceptable for high-temperature PEM fuel cells, that the cost of a fully integrated high-temperature PEM fuel cell system should always be less than a comparably sized low-temperature PEM fuel cell system. However the analysis is complicated by the fact that the entire system cost must be considered, and there are many differences at the system level.

For example, high-temperature PEM fuel cell stacks are more expensive than low-temperature PEM fuel cell stacks (by as much as 2 times). High-temperature PEM fuel cells do not require a cathode-air humidifier; low-temperature PEM fuel cells do. The entire fuel cell system and balance-of-plant must be comparatively analyzed.

With respect to the methanol reformer, if a palladium-zinc reforming catalyst is used and the operating conditions are such that the carbon monoxide level in the reformate stream is ≤3%, no further purification is necessary. This approach to reforming would be less expensive than if hydrogen purification must be included.

If a more conventional copper-zinc or platinum catalyst is used, then some degree of hydrogen purification is required. The WGS reaction would be the most likely chemical purification step, and this has been discussed above. In order to achieve sufficiently low carbon monoxide concentrations, excess water must be present with methanol in the reforming reactor—the preferred 1:1 molar ratio of methanol to water (64 wt% methanol) will not yield acceptable results. Exactly how much excess water must be present can be calculated from the WGS equilibrium expression, the exit temperature of the shift reactor(s) and the target carbon monoxide concentration.

As an example a 50 wt% methanol in water mix (1 gmol methanol and 1.78 gmol water) will yield about 2% carbon monoxide in the exiting reformate stream at 200°C using commercial WGS low-temperature shift catalysts. Although this result is fine for a high-temperature PEM fuel cell, the excess water causes a reduction in efficiency and thus an increase in operating cost (fuel consumption).

Table 4.4 Comparison of the ideal (theoretical) hydrogen yield from 1 L of a 50 wt% and 64 wt% methanol solutions

Methanol Concentration	For Each Liter of Solution		
	Total Mass	Methanol Mass	Ideal Hydrogen Yield
50 wt%	916.3 g	458.2 g	42.95 gmol
64 wt%	885.5 g	566.7 g	53.13 gmol

Comparing the theoretical hydrogen yield from an equivalent volume of 50 wt% methanol in water mix and 64 wt% methanol in water shows that the water rich mixture will result in only 80% of the hydrogen that can be produced from the 64 wt% mixture (Table 4.4). Since liquid fuels are universally sold on a volumetric basis, this means that using 50 wt% methanol solution results in an operating cost penalty of 24%.

As previously discussed (section 4.2.3) both a high-temperature shift reactor and a larger low-temperature shift reactor are commonly used in series. Interstage cooling is also necessary, in part because the WGS reaction is slightly exothermic; but in any event the reformate stream must be cooled from the temperature of the reforming reactor (commonly 300°C to 400°C) to about 200°C. The cost of the shift reactors plus interstage cooling could well exceed the cost of the reforming reactor plus vaporizer.

If the application is with a small-scale fuel cell, the reformate flow rate will be low and a single variable-temperature WGS reactor may be worth considering. The term "variable temperature" reflects the fact that the WGS catalytic reactor is intentionally operated over a broad temperature range that may span 200°C between the inlet temperature and the exit temperature. Since heat must be dissipated through the wall of the variable-temperature shift reactor (or alternative means, such as an interior heat exchange loop) the reactor is likely to be relatively large and expensive.

In summary, the economic and technical merits of methanol reforming for high-temperature PEM fuel cells are dependent on many factors. It is important to understand the intended use when making fundamental choices, such as reforming catalyst composition, in order to properly weigh tradeoffs between operating cost and capital cost.

4.4 Commercial Sources and Technical Challenges

Selective oxidation reactors and selective methanation reactors are not sold commercially as a complete subsystem scaled to a given feed stream (flow rate and composition). So any interest in using these chemical purification methods should be approached as a development project. There are commercial catalyst formulations for selective oxidation of carbon monoxide (e.g., Süd-Chemie, Inc., and BASF) and they are commonly based on platinum doped with iron and supported on alumina.[25] Selective methanation catalysts are available from Süd-Chemie, and Pacific Northwest National Laboratory and Institute of gas Technology have published results of formulations made in their labs.[7,8]

Palladium-alloy hydrogen purifiers are offered by several companies including Element One, Power+Energy, REB Research and Consulting, Pall Corp., Hy9 Corp., ATI Wah Chang, CJB Wellman, Johnson Matthey, Parker Hannifin Corp., Membrane Reactor Technology Ltd., Japan Pionics, and IdaTech. Both planar and tubular membranes are used in offered products. Sealing methods include soft gaskets, brazing, solid-state diffusion bonding, and welding.

At least one company, Pall Corp., offers a hydrogen-selective membrane that is made by plating palladium and other alloying metals onto a porous tubular support. Hy9 Corp. uses an acid-etching process to reduce the as-rolled thickness of rolled Pd-40Cu membranes and a diffusion bond to make the seals. Element One and IdaTech use ultra-thin rolled palladium-copper foils and soft gasket seals.

Even fewer companies offer commercial methanol reformers that incorporate membrane-based hydrogen purification. Those that do include Element One, Genesis Fueltech, IdaTech, Hy9, and REB Research and Consulting. Since available products and pricing may vary with time, interested readers should conduct an internet search to obtain the most recent information, including contact person, for each company.

Notes

1. Ballard Power Systems, Inc., Mark1020 ACS™ Fuel Cell Stack Product Manual and Integration Guide; document number MAN5100192-0F (February 25, 2009) p. 16
2. IdaTech has designed, built, and tested PSA hydrogen purification systems for 5 kW fuel cell systems (low-temperature PEMFC), but the systems were large, complex to control, and costly and offered hydrogen yields marginally equivalent to membrane-based hydrogen purification.
3. L. Lloyd, D.E. Ridler, and M.V. Twigg, "The water-gas shift reaction," Chapter 6 in *Catalyst Handbook*, 2nd edition, edited by M.V. Twigg, Wolff Publishing, London, UK (1989), pp. 283–339
4. *Catalyst Handbook*, 2nd edition, edited by M.V. Twigg, Wolff Publishing, London, UK (1989) Appendix 7, pp. 543–548
5. R.A. Dagle, A. Platon, A.K. Datye, J.M. Vohs, Y. Wang, and D.R. Palo, "PdZnAl catalysts for the reactions of water-gas-shift, methanol steam reforming, and reverse-water-gas-shift," *Appl. Catal. A*, 342, No. 1–2 (2008) pp. 63–68
6. W. Ruettinger, O. Ilinich, and R.J. Farrauto, "Operating characteristics for two precious metal monolithic water-gas shift catalysts for hydrogen generation," poster presented at the 20th Meeting of the North American Catalysis Society, Houston, TX, June 17–22, 2007, http://www.nacatsoc/20nam/abstracts/P-S12-40B.pdf
7. W.P. Acker, *Diagnostic method and control of preferential oxidation of carbon monoxide*, US Patent #6,322,917 (November 27, 2001)
8. M.A. Meltser, and M.M. Hoch, *Controlled CO preferential oxidation*, US Patent #5,637,415 (June 10, 1997)
9. M.S. Al-Ghamdi, "Studies on catalysts for preferential oxidation of CO in H_2-rich gas mixture," master of science thesis in chemical engineering, King Fahd University of Petroleum & Minerals, Dhahran, Saudi Arabia (June 12, 2006)
10. M.J. Castaldi, R.S. Boorse, S. Roychoudhury, P.V. Menacherry, and W.C. Pfefferle, "A compact, lightweight, fast-response preferential oxidation reactor for PEM automotive fuel cell applications," technical paper posted on the Precision Combustion, Inc., website, http://www.precision-combustion.com/proxpaper.pdf, accessed on September 28, 2009
11. B.S. Baker, J. Huebler, H.R. Linden, amd J. Meek, *Process for selective removal by methanation of carbon monoxide from a mixture of gases containing carbon dioxide*, US Patent #3,615,164 (October 26, 1971)
12. A. Rehmat and S.S. Randhava, "Selective Methanation of Carbon Monoxide," *Ind. Eng. Chem. Prod. Res. Dev.* 9, No. 4 (1970) pp. 512–515
13. R.A. Dagle, Y. Wang, G.-G. Xia, J.J. Strohm, J. Holladay, and D.R. Palo, "Selective CO methanation catalysts for fuel processing applications," *Appl. Catal. A*, 326, No. 2 (2007) pp. 213–218
14. D. Edlund, "Hydrogen membrane technologies and application in fuel processing," Chapter 8 in *Hydrogen and Syngas Production and Purification Technologies*, edited by K. Liu, C. Song, and V. Subramani, John Wiley & Sons, Hoboken, New Jersey (2010) pp. 357–384
15. D. Edlund, "Engineering scale-up for hydrogen transport membranes," Chapter 5 in *Nonporous Inorganic Membranes for Chemical Processing*, edited by A.F. Sammells and M.V. Mundschau, Wiley VCH, Weinheim, Germany (2006) pp. 139–164

16. S.N. Paglieri, "Palladium Membranes," Chapter 3 in *Nonporous Inorganic Membranes for Chemical Processing*, edited by A.F. Sammells and M.V. Mundschau, Wiley VCH, Weinheim, Germany (2006) pp. 77–106
17. H. Conner, "Palladium alloy diffusion cells," *Platinum Metals Rev.*, 6, No. 4 (1962) pp. 130–135
18. A.G. Knapton, "Palladium alloys for hydrogen diffusion membranes," *Platinum Metals Rev.*, 21, No. 2 (1977) pp. 44–50
19. F.A. Lewis, "The palladium-hydrogen system," *Platinum Metals Rev.*, 26, No. 1 (1982) pp. 20–27
20. http://www.hydrogen.energy.gov/annual_review09_production.html, accessed on January 25, 2010
21. P.M. Roberts and D.A. Stiles, "Palladium alloy diffusion units," *Platinum Metals Rev.*, 13, No. 4 (1969) pp. 141–145
22. R. Hughes, "Composite palladium membranes for catalytic membrane reactors," *Membr. Technol.*, 13, No. 3 (2001) pp. 9–13
23. D.J. Edlund, C.R. Hill, W.A. Pledger, and R.T Studebaker, *Hydrogen purification devices, components and fuel processing systems containing the same*, US Patent #6,953,497 (October 11, 2005)
24. D.J. Edlund and W.A. Pledger, *Hydrogen producing fuel processing system*, US Patent #6,221,117 (April 24, 2001)
25. O. Korotkikh, R.J. Farrauto, and A. McFarland, *Method for preparation of catalytic material for selective oxidation and catalyst members thereof*, US Patent #6,559,094 (May 6, 2003)

Chapter 5

MEMBRANE REACTORS FOR METHANOL REFORMING

Membrane reactors are a combination of a chemical reactor and a membrane separation device. The membrane should be permselective for one of the products of the chemical reaction *and* the membrane must be in the vicinity of the chemical reaction. The goal is to drive an equilibrium-limited reaction toward complete conversion by selectively removing a product from the reaction zone. For high-temperature reforming reactions, polymeric membranes are not suitable, but inorganic membranes are chemically and thermally stable under the reaction conditions. Many reports have been published describing the effective use of membrane reactors for a variety of high-temperature reactions, including methanol-steam reforming.[1–3]

Palladium-alloy membranes are a good choice for use in a membrane reactor if hydrogen is a principal product. Alternatively, a thermally and chemically stable membrane that is permselective for carbon dioxide would also be a good choice, although at the present no such membranes are commercial available.

Favorable attributes of membrane reactors include the potential for a smaller and less costly overall system by virtue of combining reaction and separation into one unit operation. One may also expect the forward reaction rates will be enhanced (since reverse reaction rates are necessarily decreased), thereby further reducing

Methanol Fuel Cell Systems: Advancing Towards Commercialization
By Dave Edlund

www.panstanford.com

the size of the reactor. Also, conversion beyond normal equilibrium limitations can be achieved in a single pass, reducing the requirement for recycling unconverted feedstock.

Drawbacks include a very challenging design that may drive cost up and typically poor mass transfer when catalyst and membrane are placed within the same physical space. Since methanol-steam reforming is not an equilibrium-limited reaction (the reverse reaction kinetics are slow), application of membrane reactors to methanol-steam reforming is not a recommended approach. This chapter will examine the concept of membrane reactors as applied to methanol-steam reforming and explain why this technical approach is disfavored.

Although many technical papers have been published with titles suggesting that they address the application of membrane reactors to methanol reforming, the majority of these papers really discuss the reaction of carbon monoxide with water to yield hydrogen within a membrane reactor. This is an important distinction, and the remainder of this chapter will deal first with methanol-steam reforming and then the water-gas shift reaction, and the suitability of membrane reactors for each reaction.

5.1 Reactor Performance

The goal of a membrane reactor is to draw off a product species from a chemical reactor during the course of reaction. In this case, the reaction of interest is methanol-steam reforming. The two principal reaction products are hydrogen and carbon dioxide. Carbon monoxide is not considered a principal reaction product; it is considered an intermediate product and is better converted to carbon dioxide and hydrogen via the water-gas shift reaction.

The chemical reactor must cause the generation of hydrogen and carbon dioxide. Since there are no commercial, high-temperature, carbon-dioxide-selective membranes, we will focus the discussion on a hydrogen-selective membrane. As the reaction proceeds, hydrogen must come in contact with, and diffuse through, the permselective membrane.

To understand how the membrane will perform in the context of a membrane reactor, first the performance of the reactor must

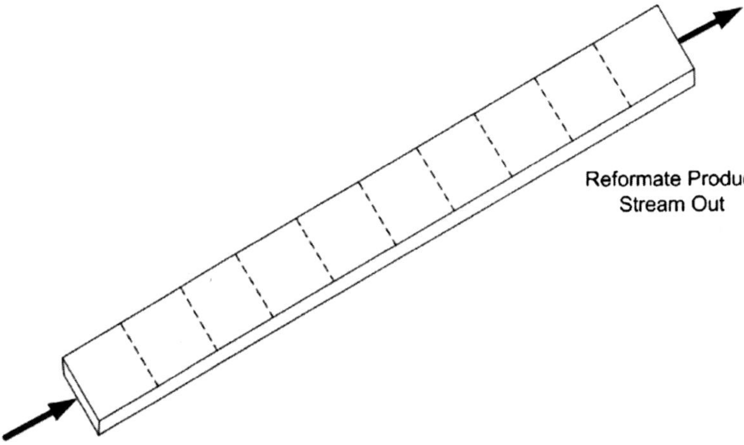

Figure 5.1 Hypothetical slab reactor for methanol-steam-reforming with segments denoted by dashed lines.

be established. As an example, consider a simple slab reactor: i.e., a rectangular reactor that is long relative to its width and height (see Fig. 5.1). It is assumed that the reactor is just large enough so that complete conversion of methanol (and water) occurs immediately prior to the exit from the reactor. It is also assumed that the conversion is directly proportional to the distance the feed stream has traveled along the length of the reactor. In other words, 50% conversion occurs when the feed stream has traveled halfway along the length of the reactor.

The example shown in Fig. 5.1 indicates 10 segments, all of equivalent length, so we may consider that 10% conversion of methanol (and water) occurs within each of the 10 segments. For this discussion it does not matter what is the operating temperature or pressure, but only that the conditions and catalyst are such that 10% conversion of the reactants occurs within each reactor segment. Also, the shape and size of the segments is not relevant; only that the assumed conversion of reactants occurs within each segment.

Making a further assumption that the feed stream is a 1:1 molar ratio of methanol and water and that the reaction proceeds according to the ideal stoichiometry ($CH_3OH + H_2O = CO_2 + 3H_2$) a plot of hydrogen partial pressure and hydrogen mole-fraction as a function

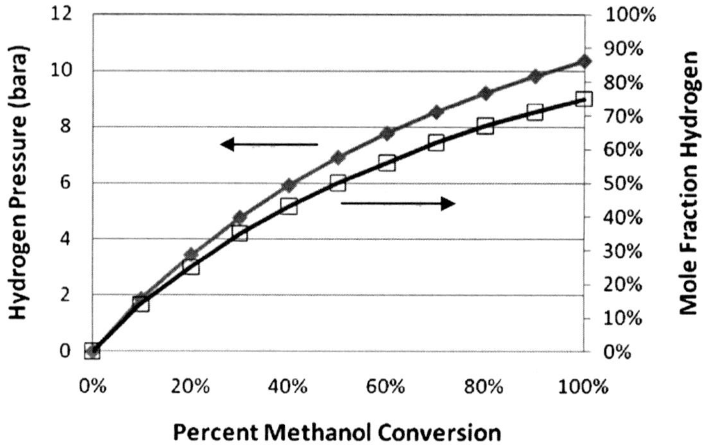

Figure 5.2 Calculated hydrogen partial pressure and mole fraction from the slab reactor shown in Fig. 5.1: total feed pressure is 13.79 bara (200 psia) and the feed stream is methanol/water mix (1:1 molar ratio).

of distance traveled along the length of the slab reactor is presented in Fig. 5.2. Initially, there is no hydrogen present and it builds up as the distance along the reactor progresses toward the end of the reactor. At complete conversion the reformate stream (product gas) is 75% hydrogen.

In a membrane reactor this reaction profile largely governs the performance of the hydrogen-selective membrane since the reactor and membrane module are intimately associated within the same vessel. This means that the hydrogen partial pressure is increasing—from initially zero—as the feed stream enters and passes through the reactor/membrane module. Although this example used a simple geometry (slab reactor), any geometry will yield conceptually similar results.

5.2 Combining Reaction with Separation

The traditional engineering approach of placing separations downstream from a reactor has served industry well for many decades. Recently, however, the concept of integrating a chemical reactor with product separation has offered the promise of an elegant

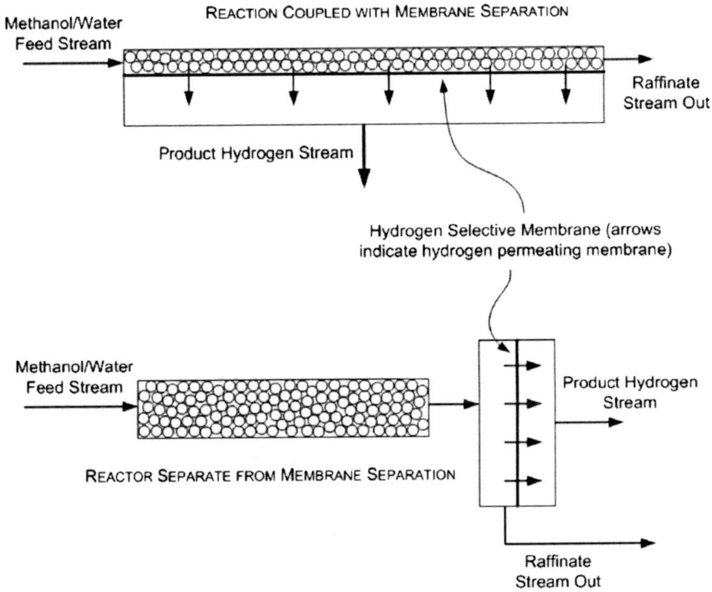

Figure 5.3 Comparison of a simple membrane-reactor design (top) and a conventional process (below) in which chemical reactor and the hydrogen-selective membrane are separate unit operations.

engineering design. A very simple visual comparison of the two approaches is shown in Fig. 5.3.

Some perceived advantages of membrane reactors include fewer process vessels and simplified piping, both contributing to a reduction in capital cost. Yet the simplistic representation in Fig. 5.3 masks many design and process details, some of which present substantial challenges—challenges that have impeded the commercialization of membrane reactors.

However, in the case of methanol-steam reforming many of the design and process details warrant careful consideration, including the following:

- Properly matching the reactor volume and reforming-catalyst area (chemical reactor) with the required membrane area (product separation)
- Securing the catalyst within the reaction region in such a way that it does not contact and damage the hydrogen-selective membrane

- Designing the reaction region so that mass transfer to the hydrogen-selective membrane is rapid
- Providing for adequate heat transfer to the reaction region (the reforming reaction is endothermic) while maintaining the hydrogen-selective membrane within the desired temperature range (avoiding overheating or underheating)

Adequately designing and controlling a membrane reactor to resolve the above issues is very challenging, resulting in costly designs and/or poor reliability. This is discussed below.

5.2.1 *Membrane Sizing*

Referring to the hypothetical slab reactor shown in Fig. 5.1, the formation of hydrogen along the length of the reactor (graphed in Fig. 5.2) provides a basis for calculating membrane performance. The results are shown in Fig. 5.4. The analysis is based on the following assumptions: total feed pressure is constant (13.79 bara) and the permeate pressure is fixed at 1.5 bara (22 psia); the feedstream is reformate derived from 1:1 methanol/water mix following ideal

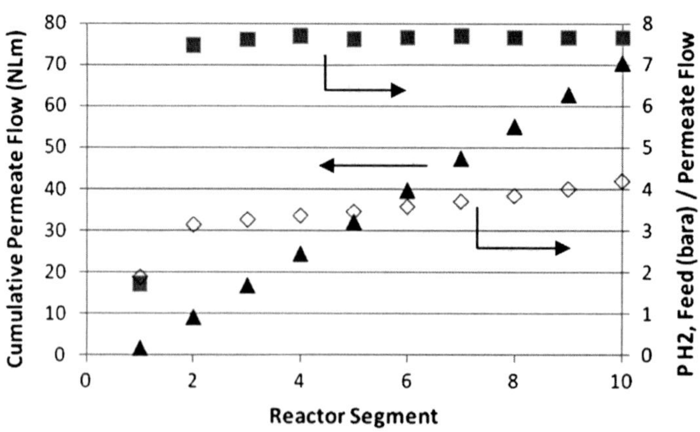

Figure 5.4 Calculated performance of a hydrogen-selective membrane in a membrane reactor based on the slab reactor shown in Fig. 5.1: the feed stream is reformate generated from methanol/water mix (1:1 molar ratio). Permeate hydrogen flow per segment (■), cumulative permeate hydrogen flow (▲), and hydrogen partial pressure in the feed stream (◇).

reaction stoichiometry; the hydrogen-selective membrane permeability is representative of commercial palladium-copper membranes, 15 microns thick; and the membrane area is allowed to vary but the hydrogen partial pressure in the raffinate stream (feed stream exiting each segment) is fixed at 11.2%.

Initially, there is no net permeation of hydrogen across the membrane since the hydrogen partial pressure remains low within the first segment of the reactor. As the graph shows, hydrogen permeation begins in the second segment, but even here the permeation rate is low. Finally, the hydrogen partial pressure reaches a high enough value by segment three that permeation is substantial, and the permeation rate then remains almost constant along the remaining length of the reactor. This analysis yields an effective hydrogen recovery of 93.9% (the theoretical maximum hydrogen in reformate is 75%) and the total required membrane area is 14,583 cm^2 to deliver 40 NLm product hydrogen.

One conclusion from this analysis is that any membrane that is placed near the entrance to the membrane reactor will be underutilized since the hydrogen partial pressure is at its lowest values in this region of the reactor. Therefore a rational membrane reactor design should not place the hydrogen-selective membrane in the first 20% or so of the reactor volume.

Another conclusion is that within the membrane reactor the hydrogen partial pressure does not rise above 4.2 bara and, over most of the length of the reactor and membrane the hydrogen partial pressure is within the range of 3 to 4.2 bara. The reason is that once the hydrogen partial pressure within the reaction zone at the feed side of the membrane exceeds the permeate hydrogen pressure; there is a net permeation of hydrogen across the membrane. As hydrogen is withdrawn from the feed side of the membrane its partial pressure is reduced. Indeed, this is why membrane reactors can shift a reaction toward more complete conversion.

In comparison, in a conventional approach wherein the reactor and membrane module are separate unit operations (bottom drawing, Fig. 5.3), the hydrogen partial pressure would be at its maximum value as the feed stream enters the membrane module. Using the above example, the hydrogen partial pressure exiting the reactor would be 10.34 bara and this would be the hydrogen partial pressure in the feed stream entering the membrane module. The hydrogen partial pressure over the feed side of the membrane in the conventional approach is much higher than in the membrane reactor concept, leading to more effective utilization of hydrogen-selective membrane.

For the same conditions used for the calculations in Fig. 5.4, the membrane area for a conventional reactor plus separate membrane module would be 1,043 cm^2; less than one-tenth the required membrane area for the membrane reactor design. Since the hydrogen-separation membrane module is relatively expensive (compared with the methanol-steam reforming reactor), increasing the size of the hydrogen-selective membrane module is counterproductive if cost is an important commercial driver.

5.2.2 *Designing the Reaction Region of a Membrane Reactor*

Although Fig. 5.3 depicts a pellet or granular form of reforming catalyst adjacent to the hydrogen-selective membrane, in fact this concept is fraught with problems because the hydrogen-selective membrane is often a thin metal foil. Catalyst pellets or granules—essentially ceramic supports with a catalytically active metal(s) layer on the surface of the particle—are highly abrasive and can easily tear or puncture the thin foil membrane due to settling, thermal expansion/contraction cycles, or shock and vibration.

Separating a conventional packed-bed catalyst from the hydrogen-selective membrane requires a physical barrier that in itself will not damage the foil membrane and yet will retain the catalyst from contacting the membrane while offering little mass transfer resistance to the membrane surface. A fine-mesh screen could suffice as such a barrier.

Alternatively, the catalyst may be coated onto a substrate; for example a corrugated foil or a metal foam or mesh. Supported catalysts of this design may be placed within the reaction cavity while avoiding contact with the membrane surface.

Regardless how the catalyst is placed with the reaction region, it is imperative that the reaction region have very little depth above the hydrogen-selective membrane. The reason is due to mass-transfer resistance. As reaction occurs at the catalyst, product hydrogen must quickly come in contact with the surface of the membrane. Failure to do so means that the product hydrogen simply passes out of the reactor and is never separated from the reaction products. As the depth of the reaction zone above the hydrogen-selective membrane increases, it becomes less likely that all product hydrogen will contact the membrane prior to being expelled from the reactor.

Fortunately, designing a thin reaction region above the membrane surface has the added benefit of facilitating heat transfer into the reaction zone. This is discussed in more detail below.

5.2.3 *Thermal Management*

Given that methanol-steam reforming is an endothermic reaction ($\Delta H°298$ = +49.3 kJ/mol), heat must be rapidly transferred into the reaction zone. In the slab reactor shown in Fig. 5.1 heat can be transferred into the reaction region from all four external surfaces. However, if the slab reactor is joined to a membrane to make a membrane reactor, the face of the reactor adjoining the membrane is effectively eliminated as an interface for heat transfer.

Planar membrane modules are usually built by stacking multiple membranes together to yield feed channels bounded on both sides by hydrogen-selective membranes. But this membrane module design does not lend itself to adaptation to a membrane reactor by placing reforming catalyst into the feed channels since heat transfer into the reaction zone is blocked by the presence of the membrane layers.

In fact, the heat transfer requirement practically limits reasonable geometric options for a membrane reactor to a single reaction zone located adjacent to a single hydrogen-selective membrane, or two reaction zones located on either side of a pair of hydrogen-selective membranes arranged such that each membrane is on either side of a central permeate collection region. Both of these options suggest large membranes and large, thin reaction regions.

Tubular hydrogen-selective membranes suffer from similar design challenges. It is not practical to place catalyst within the bore of small-diameter tubes, and placing the catalyst around the outside of the tube (typically a tube bundle) presents challenges for efficient heat transfer and mass transfer. If larger-diameter membrane tubes are employed (approximately 1.5 cm diameter or larger), then packing the bore with catalyst is feasible, but the same challenges with respect to heat transfer and mass transfer are still present.

It is not practical to superheat the incoming methanol/water feed stream to provide the heat needed for reforming. The feed stream would have to be superheated to a temperature more than 400ºC above the minimum desired reforming temperature. Since the reforming temperature and membrane temperature are about the same in the membrane reactor, and since the membrane performance (at least for Pd-40Cu membranes) is optimal within the temperature range of 350°C to 400°C, the feed stream would have to be super-heated to at least 750°C, perhaps 800°C. This is very hot and imposes expensive material requirements.

Even if the cost of materials was acceptable for supplying a superheated feed stream to the reaction zone, regulating the operating temperature of the membrane is a difficult challenge. Initially the feed entering the membrane reactor would be at a high temperature and cause local overheating of the hydrogen-selective membrane. At best the membrane permeability would be substantially reduced; at worst the membrane would fail from the excessive temperature. As the reforming reaction occurs, the temperature of the feed stream would be incrementally reduced, but only reaching the preferred temperature in the last 20% to 30% of the reactor.

5.3 Conclusion and Potential for Commercial Success

Partly because the methanol-steam reforming reaction is not equilibrium limited, the prognosis for a commercially successful membrane reactor is poor. A conventional packed-bed catalytic reactor using commercial platinum-based catalysts and operating at 350°C to 400°C will yield >95% conversion at a gas-hourly space velocity of 3,000 to 4,000 hour^{-1}. This means that reforming reactors based on conventional design are small and low cost.

In contrast, membrane reactors for reforming methanol/ water mix require more than 10 times more membrane area and the consequent increase in cost is substantial. Additionally, designing a membrane reactor so that heat transfer into the reactor is practical suggests a geometric design that is relatively large. The need to accommodate a large membrane area adjacent to the reaction zone also leads to a relatively large design.

Finally, the requirement to protect the thin foil membrane from damage due to contact with the reforming catalyst suggests either a physical barrier to separate the catalyst from the membrane (such as a fine metal mesh) or a supported catalyst. Examples of supported catalyst structures include corrugated metal foil, metal mesh, or metal or ceramic foam that is coated with a thin layer of the reforming catalyst. These supported catalysts would then need to be inserted in the reaction zone and fixed in place to avoid contact with the membrane surface.

The above discussion of the drawbacks of using a membrane reactor for facilitating methanol-steam reforming does not necessarily apply to the water-gas shift reaction. Since carbon monoxide is a by-product of methanol reforming, membrane reactors may be applied to drive further conversion of carbon monoxide to hydrogen. Although this is not the same as applying membrane reactor technology to methanol reforming, the end result is arguably the same; that is, increasing production of hydrogen. A reactor design proposed by Buxbaum[1] includes a methanol-steam reformer at the front end (e.g., a packed-bed catalytic reactor) followed by a water-gas shift membrane reactor.

The water-gas shift reaction is well suited for membrane reactors incorporating a hydrogen-selective membrane. This is because the reaction is exothermic ($\Delta H°298 = -41.4$ kJ/mol) and is subject to equilibrium control, meaning that both the forward and reverse reaction rates are fast in the presence of a catalyst. There have been many published papers describing the effectiveness of a membrane reactor for driving the water-gas shift reaction toward complete conversion by selectively removing hydrogen from the reaction zone.[4–8] A comprehensive discussion of membrane reactors using palladium-alloy membranes as applied to the water-gas shift reaction is provided by A. Criscuoli.[9]

Although heat transfer into the water-gas shift membrane reactor is not required, heat must be dissipated from the membrane reactor to avoid overheating the membrane. The operating temperature is not much of a concern for the catalytic reaction since the reaction is driven toward complete conversion by removal of product hydrogen. Still, interstage cooling or some other means of removing heat from the membrane reactor must be economically designed into the membrane reactor. Also, the challenges previously discussed concerning placing catalyst in the immediate vicinity of the hydrogen-selective membrane are just as valid for the water-gas shift reaction as for methanol-steam reforming. Nevertheless, at least one study[10] concludes that water-gas-shift membrane reactors can offer cost advantages over more conventional approaches.

Notes

1. R.E. Buxbaum, "Membrane reactor advantages for methanol reforming and similar reactions," *Sep. Sci. Technol.*, 34, No. 10 (1999) pp. 2113–2123
2. D.-W. Lee, S.-J. Park, C.-Y. Yu, S.-K. Ihm, and K.-H. Lee, "Study on methanol reforming-inorganic membrane reactors combined with water-gas shift reaction and relationship between membrane performance and methanol conversion," *J. Membr. Sci.*, 316, No. 1–2 (2008) pp. 63–72
3. C.-H. Fu and J.C.S. Wu, "A transient study of double-jacketed membrane reactor via methanol steam reforming," *Int. J. Hydrogen Energy*, 33, No. 24 (2009) pp. 7435–7443, doi:10.1016/j.ijhydene.2008.10.005
4. D. Edlund, D. Friesen, B. Johnson, and W. Pledger, "Hydrogen-permeable metal membranes for high-temperature gas separations," *Gas Sep. Purif.*, 8, No. 3 (1994) pp. 131–136
5. A. Basile, A. Criscuoli, F. Santella, and E. Drioli, "Membrane reactor for water-gas shift reaction," *Gas Sep. Purif.*, 10, No. 4 (1996) pp. 243–254
6. S. Uemiya, N. Sato, and H. Ando, "The water-gas shift reaction assisted by a palladium membrane reactor," *Ind. Eng. Chem. Res.*, 30, No. 3 (1991) pp. 585–589
7. A. Criscuoli, A. Basile, and E. Drioli, "An analysis of the performance of membrane reactors for the water-gas shift reaction using gas feed mixtures," *Catalysis Today*, 56, No. 1–3 (February 25, 2000) pp. 53–64
8. S. Tosti, A. Basile, G. Chiappetta, C. Rizzello, and V. Violante, "Pd-Ag membrane reactors for water-gas shift reaction," *Chem. Eng. J.*, 93, No. 1 (2003) pp. 23–30
9. A. Criscuoli, "Economics associated with implementation of membrane reactors," in *Nonporous Inorganic Membranes*, edited by A.F. Sammells and M.V. Mundschau, Wiley-VCH, WeinHeim, Germany (2006) pp. 245–263
10. A. Criscuoli, A. Basile, E. Drioli, and O. Loiacono, "An economic feasibility study for water-gas-shift membrane reactor," *J. Memb. Sci.*, 181, No. 1 (2001) pp. 21–27

Chapter 6

BARRIERS TO COMMERCIALIZATION

A handful of companies are actively developing commercial fuel cell systems incorporating a methanol reformer. IdaTech and Protonex have publicly introduced either commercial products or prototypes. Previously several of the global automakers also developed concept cars that incorporated a methanol reformer and low-temperature PEM fuel cells. Daimler Benz road tested several generations of concept car—called Necar—and Toyota also road tested a version of the Rav4. All of the vehicle activity using onboard reforming has come to an end; the automakers appear to favor compressed hydrogen storage due to simplicity, cost reduction, and rapid startup.

There is a growing recognition that methanol is a versatile fuel and that reformers producing pure hydrogen have fundamental and profound advantages over compressed hydrogen when coupled to fuel cells for generating electricity. This chapter will address the status of commercial products and development efforts.

6.1 Commercial Status: Reformed-Methanol Fuel Cell Systems

Perhaps the two companies best known for reformed-methanol fuel cell systems are IdaTech and Protonex Technology Corp. Others working in the field to develop at least prototype products include Chung-Hsin Electric & Machinery Mfg. Co., Intelligent

Methanol Fuel Cell Systems: Advancing Towards Commercialization
By Dave Edlund

www.panstanford.com

Energy, Altergy Systems, Casio, Dantherm Power, ReliOn and Ballard Power Systems.

IdaTech has achieved moderate commercial success with their ElectraGen™3 and ElectraGen™5 products—rated at 3 kW and 5 kW, respectively—that are designed to address grid-tied backup power applications. The most significant market now is global telecom, where regulations and customer satisfaction have driven the industry to provide service even in the event of grid power failure.

The ElectraGen family uses a Ballard air-cooled PEM fuel cell. The methanol reformer is an internally developed methanol-steam reformer and it is directly coupled to a hydrogen purifier using a Pd-40Cu planar foil membrane. The hydrogen purification module also is an internally developed design. The reformers are required to produce about 13 NLm of purified hydrogen per kW electric output from the fuel cell stack. A consequence of using the membrane-based hydrogen purification is that the reformer must operate at elevated pressure, typically between 100 psig and 150 psig. Since the target application is stationary backup power, size and weight are not important criteria.

Figure 6.1 ElectraGen XTR reformed methanol fuel cell system sold by IdaTech for use as a backup power system, primarily for grid-tied telecommunications applications (photo courtesy of IdaTech, LLC). See also Color Insert.

Figure 6.2 The M250 product family under development by Protonex Technology Corporation and aimed at the recreational market (photo courtesy of Protonex Technology Corp.). See also Color Insert.

Protonex Technology Corp. has invested in developing a man-portable 250 W fuel cell system that uses a compact and light-weight methanol reformer to supply pure hydrogen to a high-power density PEM fuel cell. Initially funded by the US Department of Defense, this development effort has resulted in working prototypes (Fig. 6.2) that have been purchased by the US Army for test and evaluation.

Protonex is developing two versions of the M250 reformed-methanol fuel cell system: the M250-B and the M250-C. Internally the two products are very similar. The principal difference is that the M250-B is designed to be a portable fuel cell system that will provide both 12 VDC and 120 VAC electrical power. The M250-C is designed to be installed within an RV or boat and it will continuously monitor the state of charge of the house lead-acid battery bank. When the batteries reach a defined state of discharge, the fuel cell system automatically turns on to charge the batteries.

Unlike the backup power market, size and weight are extremely important for portable fuel cell systems and the Protonex product demonstrates this understanding. Both the fuel cell stack and methanol-steam reformer are internally designed and developed. The stack (low-temperature PEM technology) is liquid cooled and operates at a peak design current density of about 700 mA/cm^2. Protonex uses a patented injection-molded seal that encases the stack in silicone rubber and seals the gas manifolding and MEAs at the same time.

Providing pure hydrogen to the PEM stack is the methanol-steam reformer coupled to a membrane-based purifier using palladium-alloy foil membranes; this comprises the fuel processor module. Although this sounds similar to the ElectraGen family, the similarities do not go very deep. To achieve light weight and rapid cold-start capability, the Protonex reformer and purifier are electrically heated using aluminum as a thermal conduction medium and heat transfer medium. Electric cartridge heaters heat a mass of aluminum, appropriately shaped to contact the significant surfaces of the reformer and purifier. With this approach a cold start time of 20 minutes can be achieved, which is very good for a portable reformer of this size.

The high degree of system integration is evident in the detailed assembly diagram (Fig. 6.3) for a M250 portable fuel cell system. The reformer needs to deliver 4 NLm of purified hydrogen to the fuel cell to net 250 W electrical output from the system. The relatively low flow rates are amenable to small internal fluid passages in the reformer and hydrogen purifier, and this is the design approach that leads to the most compact size and minimum weight. But small fluid

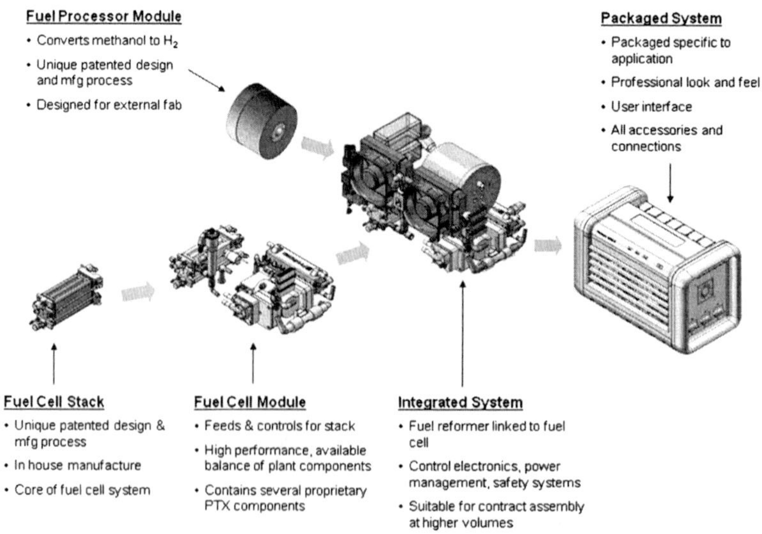

Figure 6.3 The M250 product family under development by Protonex Technology Corporation and aimed at the recreational market (figure courtesy of Protonex Technology Corp.). See also Color Insert.

passages are prone to plugging from debris (including dust from the reforming catalyst and non-volatile material that originates in the methanol/water feed solution) and care must be exercised to ensure proper design for uniform flow.

6.2 Commercial Status: Methanol Reformer Subsystems

There is a growing commercial demand by fuel-cell system integrators for methanol reformers. Several companies have recognized the business opportunity and have developed methanol reformer products ranging from <1 L/min to 75 L/min hydrogen output. All of these products are based on methanol-steam reforming. Most of these products utilize a palladium-alloy hydrogen purifier.

REB Research and Consulting has been selling stand-alone methanol reformers as hydrogen generators for many years. Incorporating a proprietary tubular palladium-alloy hydrogen purifier, the REB products have been sold to laboratories and semiconductor manufacturers. The tube is made by drawing the palladium-alloy through a die. An example of a large methanol reformer for manufacturing operations is shown in Fig. 6.4 and a smaller table-top version for lab use is shown in Fig. 6.5.

Figure 6.4 A large methanol reformer rated to deliver 75 sLm product hydrogen at 99.99995% purity. Founder Robert Buxbaum is shown to the right (photo courtesy of REB Research and Consulting). See also Color Insert.

Figure 6.5 Bench-top methanol reformer rated to deliver 0.5 L/min high-purity product hydrogen (photo courtesy of REB Research and Consulting). See also Color Insert.

The company lists five different models on their website, ranging in hydrogen output from 0.5 sLm to 75 sLm. All models are claimed to be automatically load following and capable of delivering product hydrogen at up to 40 psig. Since the pressure of the product hydrogen permeating the palladium-alloy hydrogen purifier cannot exceed the partial pressure of hydrogen in the raffinate stream exiting the hydrogen purifier, it stands to reason that the REB methanol reformers operate with a feed pressure of at least 100 psig.

Another company that has been developing methanol reformers as stand-alone systems for many years is Genesis Fueltech. Also incorporating palladium-alloy hydrogen purification, Genesis Fueltech products are sized to deliver from 4 L/min to 20 L/min product hydrogen at 99.9999% purity.

One company in Taiwan is advertising their methanol reformer product line; Green Hydrotec Inc. Unlike all other competitors in this commercial space, Green Hydrotec offers two reformer models that appear to be specifically designed for high-temperature PEM fuel cell systems. The GHM-200LPH is rated to deliver 200 L/hour product hydrogen and the GHM-3000LPH is rated to deliver 3,000 L/hour product hydrogen. Hydrogen purification is not included with these models and the product hydrogen purity is specified to be 75% with carbon monoxide concentration not to exceed 0.5%. No mention is made in the product specification sheet as to the

Figure 6.6 Model GHM-1000LPH methanol reformer with palladium hydrogen purifier. (photo courtesy of Green Hydrotec Inc.). See also Color Insert.

concentration of water in the methanol-water blend consumed by these reformers.

Green Hydrotec also offers two models that incorporate palladium-alloy hydrogen purifiers—the GHM-500LPH and the GHM-1000LPH—rated to deliver 500 L/hour product hydrogen and 1,000 L/hour product hydrogen, respectively. A proprietary tubular palladium-alloy purifier module yields product hydrogen that is 99.99% pure, according to published specifications. Interestingly, these reformers are rare examples of commercially available products that utilize hydrogen-purification membranes made by deposition of palladium onto a porous stainless steel tube (Fig. 6.7). This approach is

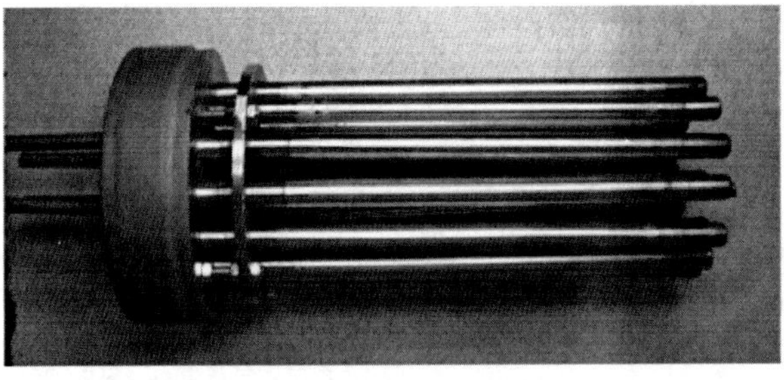

Figure 6.7 Tubular hydrogen-separation membrane bundle made by plating palladium and copper onto sintered porous stainless steel tubes. (photo courtesy of Green Hydrotec Inc.). See also Color Insert.

Table 6.1 Three methanol reformer products sold by Element One

Model	Rated Hydrogen Output	Rated Minimum Hydrogen Purity	Equivalent Fuel Cell Output
Me-10	14 NLm*	>99.95%; <1 ppm carbon monoxide	1 kW_e
Me-20	25 NLm	>99.95%; <1 ppm carbon monoxide	2 kW_e
Me-50	65 NLm	>99.95%; <1 ppm carbon monoxide	5 kW_e

* NLm is normal liters per minute and is defined at 21°C and sLm is standard liters per minute and is defined at 0°C.

not favored commercially due to high inherent cost (the cost of the porous stainless steel support tube can easily exceed the cost of a palladium-alloy foil with ten times the active area) and historically poor durability.

Likewise, methanol reformer products offered by Element One utilize high-pressure methanol-steam reforming and a membrane-based hydrogen purifier using palladium-copper foil membranes. Targeting backup power applications, three products are offered (see Table 6.1). Element One has focused on an industrial product that is the lowest priced methanol reformer system commercially available.

To achieve very low cost of manufacture, Element One has simplified the operation of the reformer eliminating load following which is costly to control and may lead to reliability issues. Indeed, the only operational data that is necessarily monitored to automatically control Element One's products is the temperature of the reformer—no feedback control over the flow rate of the methanol/water feedstock is required.

Element One's reformer products are designed to sit in a hot standby mode wherein electric power (from the grid) is used to energize electric cartridge heaters and maintain the reformer and hydrogen purifier at between 300°C and 350°C. At the moment that grid power is lost, the methanol reformer system automatically turns on and uses combustion to maintain the reformer at an operating temperature between 350°C and 400°C. Hot reformate from the reformer heats the hydrogen purifier internally, so no combustion gases are allowed to flow over the hydrogen purifier module.

Hy9 Corporation recently launched two models of their methanol reformer: the HGS-M-40 is rated to deliver 40 sLm product hydrogen and the HGS-M-80 is rated to deliver 80 sLm product hydrogen. These products are clearly aimed at backup power applications. Both models use an internally developed palladium-copper hydrogen purifier. The foil membranes are approximately 20 μm thick and are made by chemically etching thicker foil to the final thickness. Seals within the membrane module are achieved by solid-state diffusion bonding. Product hydrogen quality is rated at 99.9999+% purity.

Diffusion bonding requires relatively soft metals and metal surfaces free of interfering oxide layers (or other impurities such as dirt and oil). Copper is an ideal metal for diffusion bonding, but it has poor mechanical strength especially at the operating temperature of methanol reformers. Hy9 has solved this problem by making the module structural components from copper-plated steel. However, the additional step of copper plating does add significant cost to the product.

Hy9 has extensive experience producing palladium-copper hydrogen purifiers for industrial customers (mostly electrolyzers), but methanol reformers is a recent addition to their product line. Like the Element One methanol reformers, Hy9 has chosen to use a hot-standby mode and an operating mode to greatly reduce the startup time in the event of a power loss. Generally the product specifications are very similar to those for Element One reformer products, except that the Hy9 products are larger in size and consume substantially more electrical power in operating mode.

6.2.1 *Prognosis for Further Development*

Methanol is gaining acceptance as a useful and often preferred fuel for fuel cell applications. This is in stark contrast to the general customer atmosphere prior to about 2005. Back then, fear over the toxicity of methanol dominated opinion about commercial usefulness of methanol and methanol reformers for producing hydrogen. Why this change in opinion occurred is hard to say, but the overwhelming success of SFC Energy AG in placing more than 20,000 direct methanol fuel cell systems into the recreational vehicle and boating market likely contributed to a growing acceptance of methanol as a fuel.

The cost of methanol reformers, including hydrogen purification, remains the largest barrier to widespread commercialization of products using the technology. There are many reasons why it has been difficult to reduce the cost of reformers and hydrogen purifiers, including the following:

- *Materials.* Some of the required materials (e.g., palladium alloys, platinum, and ruthenium) are inherently expensive. Failing to minimize the use of these materials, or substituting less expensive alternatives, is an impediment to realizing low cost of manufacture.
- *System design.* Simply stated, complex designs invariably lead to high manufacturing cost. The complexity of a design may be measured by the number of discrete components or pieces that must be assembled to make the reformer or purifier, the type and amount of instrumentation needed to control the reformer and purifier (e.g., thermocouples and flow sensors), and the type and amount of machining, welding, brazing, etching, or other fabrication method.
- *Operational control.* The more complex the control architecture, the more expensive the product will be. Control complexity is often manifested in multi-step startup routines and shutdown routines, and in load following whereby the reformer automatically attempts to match the hydrogen output with the hydrogen consumption of the fuel cell stack.

Still, the prognosis for successful commercialization of methanol reformers combined with fuel cell systems is good. The preceding discussion has highlighted emerging product sales of reformed-methanol fuel cell systems as well as methanol reformers that are being marketed as a subsystem for integration into fuel cell systems. It is clear that the favored technical approach is methanol-steam reforming combined with a palladium-alloy hydrogen purifier.

Although there is commonality in the technical approach, there is little consensus regarding the target applications that will be commercially successful. Backup power systems that will automatically turn on and support critical electrical loads when the grid fails are one important application, but others include military portable power, recreational battery charging on boats and RVs, and even small-scale combined heat and power applications (although it is doubtful that methanol would be a customer-accepted fuel for combined heat and power applications).

There may even be unanticipated new applications that emerge once reformed-methanol fuel cell systems become more widely recognized and the advantages understood. Possible new applications include robotics, emergency power for building and border security, and hybridized with batteries and PV for remote data collection and transmission (such as monitoring oil and gas pipelines, electric transmission lines, weather and geological monitoring sites, highway and railroad safety monitoring, animal migration monitoring, and other remote scientific data monitoring sites).

All of these applications are made possible primarily because of the ease of handling and storing methanol-water blends; the fact that the mixture will not freeze, and the low environmental impact of accidental spills.

6.3 Systemwide Economic Analysis

In this section we will examine the cost of reformed-methanol fuel cell systems and the factors that strongly influence cost. Since the cost of reformers, hydrogen purifiers, and fuel cell systems has been a historical barrier to broad commercial adoption, our goal is to understand why this has been the case and what steps might be taken to overcome the cost barriers. The four major subsystems will each be evaluated separately.

Only capital cost (or initial cost) will be considered in any detail. However, it should be remembered that in some cases total lifecycle cost is a more important figure for the customer. Lifecycle cost is more difficult to assess because the operational lifetime of the product must be known as well as the frequency and cost of maintenance, cost of methanol-water fuel, the period of time over which the product purchase price (plus applicable interest) is amortized, and the residual value (if any) at end of life.

6.3.1 *Methanol Reformer Subsystem*

The methanol reformer and associated hydrogen purifier have been discussed in detail in preceding chapters. The point of the discussion here is to emphasize critical factors that influence the cost of this important subsystem.

It should go without saying that the reformer (and fuel cell) should be sized for the real load requirements of the application, yet it is surprising how often this vital information is not know with certainty. Nameplate power ratings for equipment are not a good indicator of actual electrical draw, and yet this is often the only information available to define the power requirements.

The reformer and hydrogen purifier should emphasize simplicity of design and use commercial materials (such as common pipe and tube dimensions)—customized forms that require special mill runs or machining should never be incorporated into the design. Likewise, commercially available catalysts should be used. Catalysts may be either pellets (or granular) or coatings applied to ceramic of metal substrates. In any case, the design of the reformer should be such that it is amenable to the selected commercial catalyst physical form.

As much as possible compression fittings should be avoided due to their expense (ranging from $10 for a simple union coupling to about $50 for a complex tee). Compression fittings are also unreliable and may leak after repeated thermal cycling or if improperly installed. Welding is far better.

An igniter of some type is needed if a normal flame burner is used. Generally there are two types of igniters—hot surface igniter and spark igniter. Hot surface igniters consist of an electric resistance element that is heated to a red heat and is in contact with fuel and air to cause ignition. These igniters are extremely reliable provided that the electric resistance element is not overheated by the burner flame. Spark igniters are also reliable if installed such that the electrode is not overheated, but these igniters cause significant electromagnetic interference with microprocessors (such as in the control system). This EMI is both emitted and conducted and so is not easily shielded. For this reason, hot surface igniters are often a better and lower-cost solution.

Catalytic burners are another option (as opposed to a flame burner) and do not normally require an igniter. But combustion at a catalyst can only be locally contained at the catalyst if the fuel to air mixture is outside the boundaries for normal combustion. This is a condition that is not easily guaranteed in a reformer. If the fuel-to-air ratio drifts into that for normal combustion, then flashback occurs and the flame propagates to the point at which fuel and air are mixed, often with undesired consequences.

If membrane-based hydrogen purification is used, the reformer and purifier need to be designed with consideration to the balance-of-plant components too. Operating pressure is the single most important operating parameter: increased pressure means that the reformer and purifier can be smaller (and lower cost), but at the expense of a more costly methanol-water pump. The optimum pressure should be defined as that at which the combined cost of the pump, reformer, and purifier is at a minimum value.

6.3.2 *Fuel Cell Subsystem*

The cost of the fuel cell subsystem is dominated by the cost of the fuel cell stack, but the balance-of-plant may also represent a significant fraction of the total cost. The fuel cell stack cost is a direct result of size and power density at the operating point. The larger the stack, the higher the cost, the higher the power density, the lower the cost (on a power-normalized basis). To quantify fuel cell stack costs it is necessary to have a specific stack design. The stack designer should strive for minimizing wasted MEA area by minimizing seal area and intrusion of gas manifolds into the MEA.

Stack cost is not a linear function of size, because some components of the stack (e.g., end plates and bus plates) do not change in size or quantity as the stack is enlarged by adding more electrochemical cells. However, the cost of the stack is largely a function of how many and what size MEAs and bipolar plates are used. A rough cost for PEM fuel cell stacks using molded graphite bipolar plates is $1 to $2 per watt of electrical power generated by the stack.

Proponents of metal bipolar plates claim that they are significantly lower in cost than are molded graphite bipolar plates and just as durable if properly made and assembled into a stack. Altergy Systems has enjoyed very good commercial success of its 5 kW PEM fuel cell stacks assembled using metal bipolar plates. Intelligent Energy was a pioneer in use of metal bipolar plates and still uses this approach in their high-power density PEM fuel cell stacks. Two companies selling metal bipolar plates, typically manufactured by stamping, are Dana Corporation and TreadStone Technologies.

A fuel cell stack may be air cooled (open cathode or closed cathode*) or liquid cooled, and the choice has profound impact on the total cost of the subsystem. The analysis is sufficiently complex that it must be undertaken in the context of the application. Some guidelines are as follows:

- An open-cathode air-cooled stack will have a very simple and inexpensive balance-of-plant, but the stack cannot operate at as high a power density as can be achieved and sustained with a liquid-cooled stack. Since a fuel cell normally operates at about 50% to 70% efficiency to electricity (the balance being heat) air-cooled stacks will overheat if operated at a power density much greater than 0.20 W/cm^2 to 0.25 W/cm^2.† A liquid-cooled stack can operate well at 0.6 W/cm^2 without overheating. Of course, the ambient air temperature also has a strong influence on the maximum allowable power density that can be sustained from an air-cooled stack.
- A liquid-cooled stack has a more complex and expensive balance-of-plant because of the added liquid pump (to circulate the coolant) and heat exchanger to cool the coolant. If water is

* Open cathode refers to a stack design wherein the cathode cooling air is also the cathode reactant air. Closed cathode refers to a stack design that uses discrete and separate air flows for cathode cooling and cathode reactant air.

† Measured performance for a Ballard Mk1020 air-cooled stack; April 2009.

the coolant, then stainless steel should be used for all wetted parts to avoid corrosion and contamination of the MEAs with metal ions. Non-aqueous coolants may be used with satisfactory results, and in this case lower-cost materials are satisfactory.

- Liquid-cooled FC stacks and closed-cathode air-cooled FC stacks both require a more complex design to properly manage the humidity of the cathode air. The need for cathode air humidification depends on the technical requirements of the MEA and the power density at which the stack is to be operated. Some MEAs are more robust under low-humidity conditions than are others; the manufacturer should be consulted for specific durability and performance under low-humidity conditions. Higher power density leads to higher localized heating, and as the local temperature increases the local relative humidity decreases, exacerbating the problem. Humidifiers do not need to be expensive, and simple plate-and-frame designs using low-cost water transport membranes are a good choice. High-cost membranes such as Nafion™ are not needed and should be avoided. An alternative to cathode air humidification is direct water injection (Intelligent Energy practices this approach).
- Controlling the cathode air flow rate to maintain a constant air-to-hydrogen ratio should be avoided unless there is a compelling advantage otherwise. Usually this is only an issue when the fuel cell stack is delivering less than the maximum electrical output. The drawback to operating with excessive cathode air flow is potential drying of the MEA (see above discussion). But a suitably sized humidifier may more than offset the cost of measuring anode and cathode flow rates and associated controls.
- In applications where ambient air pollution is significant, a suitable particulate and chemical air filter is required for open-cathode air-cooled stacks (e.g., the Ballard Mk1020 PEMFC stack). The drawback is that a much larger volume of air must be filter for an open-cathode stack versus either a closed-cathode air-cooled stack or a liquid-cooled stack.

Assuming that both liquid-cooled and air-cooled stack options are available, the correct choice would be that which has the lowest combined cost of stack plus balance-of-plant.

6.3.3 *Power Electronics*

Several fuel cell system integrators have also taken on the task of designing power electronics to suitably condition the output electrical power, modifying voltage, inverting to alternating current, or applying a battery charging algorithm. One might reasonably question whether this is a wise choice for a development phase company endeavoring to commercialize fuel cell systems, especially given the number of global enterprises with demonstrated capability in designing power electronics. Internally designed power electronics can cost as much or more than the fuel cell subsystem. This should be considered unacceptable and every effort should be made to reduce the cost of power electronics to not more than 50% the cost of the fuel cell subsystem.

Commercial DC voltage converters are an affordable choice for converting raw stack voltage to the voltage required by the customer, typically 12, 24, or 48 VDC. An even better choice is to design a fuel cell stack so that the output stack voltage matches the commercial requirements, or is at a value that matches commercial DC voltage converters.

AC power must be obtained by using an inverter since fuel cells are fundamentally DC generators. Nevertheless, there are many commercial choices for inverters varying in range of input DC voltage, power, efficiency, and type of alternating current waveform. The lowest-cost inverters yield a square waveform; a true sine wave is more expensive and often unnecessary. Conversion efficiencies for commercial inverters range from 85% at the low end (very low cost inverters) to 95% and the high end.

If the application can use DC power and the load is mostly constant, then it is possible to design a fuel cell stack with the proper number of cells to give the designed output DC voltage matching the application requirements. In the case of lead-acid battery charging, it is not essential to use power electronics to control the battery charge cycle. Rather, the output from the fuel cell stack can be fed directly to the battery (or bank of batteries, assuming proper match of voltage). As the batteries charge, the current draw decreases and the stack output voltage increases. This is an example of a natural fit between the DC requirements of the load (charging the battery bank) and the fuel cell operating characteristics. All that is needed is a voltage monitor that causes the fuel cell to turn on when the voltage of the battery bank falls to a prescribed value—such as 90% of nominal voltage—and to turn off the fuel cell when the voltage reaches a higher prescribed value—such as 120% of nominal voltage.

6.3.4 *Automated Controls*

This is another area where many fuel cell companies have failed to adequately control costs. The control system must provide for safe and proper control of the system, but what is proper control? In many cases the answer is more than simply producing electrical power. Many of the product performance specifications may influence the answer, including fuel efficiency at partial load, start time from cold, transient response time, performance at high altitudes, etc. Other design features that cause the cost of the controller to escalate include the user interface and remote connectivity (e.g., wireless or Internet).

A useful exercise when laying out the automated control functional requirements is to list by priority those characteristics that are important to the customer. For instance price is usually the number one priority; fuel efficiency over a range of turn down may also be a high priority, but not always. For portable fuel cell systems rapid start from cold is important. Once it is understood what is really important, then the logical approach to controlling the fuel cell system can be designed.

An example that illustrates this process is grid-tied backup power. The fuel cell system and methanol reformer must be able to start very quickly and the price must be low because the competition is batteries, possibly augmented with a generator. The desired performance features are low cost and rapid start. No value is assigned to especially small packaging or light weight—nor to fuel efficiency and turn down capability—and transient response is not even a requirement. All the system is required to do is quickly begin to generate electrical power once the grid fails and to do so at a relatively low (competitive) purchase price and be highly reliable.

The controls designer should correctly focus on a simple and robust control scheme that causes the reformer to operate at one output. The fuel cell is always load following (physical laws demand that the fuel cell produces exactly the amount of electrical energy as is consumed by the load) and any excess hydrogen can be easily disposed of. There is absolutely no practical gain in trying to control the reformer such that it attempts to follow the electrical load. Rapid startup is easily achieved by maintaining the reformer in a hot-standby state using grid power and electrical heaters. This is very simple to implement from a controls perspective and very reliable. Also, by keeping the reformer hot essential all the time the number of thermal cycles over the lifetime of the product is very low, and this yields enhanced durability of the reformer.

Another example is a portable reformed-methanol fuel cell system. For this application, much more demanding than backup power, it is important that the system starts quickly from a cold resting state. Fuel efficiency is also important because the application is often recreational use and cost of fuel is important to the customer (as well as representing weight that must be transported). Now the controls designer must devise a scheme to put a high flux of thermal energy into the reformer over a short time span at start up, and then throttle back the thermal input so as not to overheat the reformer.

Maximizing fuel efficiency implies that the reformer must be throttled down during periods of low electrical demand. This, in turn, necessitates some means for measuring electrical output (or electrical demand) and inferring hydrogen usage, and then adjusting the reformer output by turning down the feed pump. Flow measurements are notoriously expensive and prone to errors, and changing pump settings to moderate the hydrogen output from the reformer is also fraught with complexity and cost. Although it would be convenient to use a positive displacement pump and control the frequency of the pump, this may not work due to liquid slippage at the pump due to aging and wear of internal seals or check valves, or drift in the supply voltage to the pump motor. The point is that these control challenges can be overcome, but doing so comes at a price.

Like power electronics, if cost control is not maintained over the automated control system its cost can easily exceed that of the fuel cell subsystem. The best way to ensure adequate cost control is to constantly work toward maximum simplicity. A good axiom to follow is "if the customer won't pay for the feature, it isn't truly necessary." An added advantage of simplifying process controls is that system reliability invariably increases.

6.4 Concluding Remarks

The concept of maximizing simplicity cannot be overemphasized. Simplicity has two direct, significant advantages: cost reduction and reliability improvement. Unfortunately, over the past 12 years, efforts of fuel cell companies to develop and successfully commercialize products have been delayed by overly complex designs. The good news is that there is clear evidence that this trend is being corrected.

How does the system designer or project manager ensure that the work is optimized for simplicity? The answer is to understand the application and needed performance features of the product. The answer may be difficult to arrive at if direct discussions with the customer are not possible. Understanding the difference between "must have" performance features and "optional" performance features should be the goal. For those performance specifications that are optional, the willingness of the customer to pay for those features should be clearly understood prior to incorporating them into the product.

Once the technical specifications are set it is important to avoid feature creep: the common practice to add just one more function or to improve upon the minimum acceptable performance just a bit. Perhaps it is the nature of technologists, engineers, and inventors, but feature creep is a common practice in technology development organizations. The real drawback to feature creep is twofold: increase in product cost (that likely cannot be recovered in higher product price) and delayed product introduction. Ultimately, the best test of suitability of any new fuel cell product is customer feedback and adoption. The sooner this feedback can be obtained, the better.

Marketers and sales agents are advised to emphasize the real value proposition of the fuel cell product. A decade ago the fuel cell industry often claimed that fuel cell products offered advantages of reduced emissions, high energy efficiency, long lifetime, low noise, low maintenance, low cost, fuel flexibility, compact size and light weight, and other advantages. In fact, currently commercial fuel cell systems can claim maybe a couple of the aforementioned advantages, but certainly not all. Setting realistic expectations with the customer is the first step to success. Understanding the real value proposition of the product is essential.

Chapter 7

APPLICATIONS AND MARKETS

Reformed-methanol fuel cell systems are being developed for many commercial applications. More so than any other organic fuel, methanol can be easily reformed to yield high-quality hydrogen and enable the successful application of low-temperature PEM fuel cells to generate electricity. As high-temperature PEM fuel cells mature, it is likely that methanol reformers will also be coupled to these fuel cells as well.

The first portion of this book addressed the technical features and theory of operation of methanol reformers and hydrogen purification processes. This chapter will examine some of the more promising commercial applications. It should be clear that methanol reformers offer both advantages and disadvantages, and like any technology, care should be taken to make sure that the application fit is good. An example of an application where reformed-methanol fuel cell systems are not a good fit is unmanned aerial vehicles (drones) where very rapid start and light weight are critical.

7.1 Consumer Electronics

Fuel cells have been proposed as battery replacements for many consumer electronics, including cell phones, PDAs, and laptop computers. Most of the large electronic-product manufacturing

Methanol Fuel Cell Systems: Advancing Towards Commercialization
By Dave Edlund

companies have evaluated fuel cell power systems; some are still actively engaged, including Sanyo, Toshiba, Casio, and Panasonic (Matsushita).

The power draw for cell phones and PDAs is low, only a few watts; whereas for laptop computers the power draw may range as high as 50 W. Some of the important product requirements include the following:

- High overall energy efficiency—this implies low heat loss to the consumer appliance and maximum run time on a fuel cartridge.
- Compactness in size and light weight—with lithium battery technology ever improving and products becoming smaller and more powerful, the challenges surrounding small size and light weight are significant.
- Robustness and safety—must withstand dropping (while operating in the appliance) from a height of 1 m onto a hard surface without introducing any significant safety risk even if the power system fails.
- Very low noise signature and clean emissions—no noxious or toxic fumes are acceptable given the close proximity to people when the system is running. In particular, it is imperative that emissions of carbon monoxide, formaldehyde, and formic acid be sufficiently low as to present no health risk.

The product requirements do not include operation at extremes of temperature, but performance must be guaranteed at elevations of 3,000 meters or more. In essence, the reformed-methanol fuel cell system should offer all of the functional features of the existing lithium-ion battery solution, and the additional advantage of longer run time. Furthermore, fuel cell products must be cost competitive with incumbent lithium batteries. While it is true that the convenience of longer run time (than can be achieved with a battery) and lack of charger may allow for premium pricing, it is doubtful that any more than a US$100 or US$200 markup will be embraced by consumers.

It has been very challenging for PEM fuel cells to meet this collection of requirements and still be affordable: to date no manufacturer of cell phones, PDAs, and laptop computers is offering commercial

Figure 7.1 Angstrom fuel cell power supply (center and right) that fits within the battery space of an unmodified Motorola cell phone (photo courtesy of Angstrom Power). See also Color Insert.

products with a fuel cell in place of the traditional battery pack. And with battery technology constantly improving, the challenge is getting more difficult.

Perhaps the most technically advanced PEM fuel cell power system for cell phones developed yet comes from Angstrom Power Inc. Although it does not utilize a methanol reformer, the Angstrom fuel cell power system uses a low-temperature PEM fuel cell and compressed hydrogen. It is rated to generate 3 W continuously and, remarkably, it fits within the existing battery compartment of a smart cell phone (see Fig. 7.1).

Despite the significant technical breakthroughs of Angstrom Power, the fuel remains hydrogen and supplying fuel cartridges as either a compressed gas or stored in a metal hydride bed is a significant barrier to widespread adoption. However, the challenges of designing a methanol reformer for this application are tough to solve. First the reformer must be small and light weight; it must be inexpensive and reliable; it must start quickly (when the phone is turned on) and it must not emit noxious smelling exhaust, nor can the reformer reject a significant amount of heat that would either cause overheating of the circuits within the phone or cause discomfort to the user holding the phone.

Although these challenges have not been solved to date for a methanol reformer, Motorola put considerable effort into developing a methanol reformer and high-temperature fuel cell system for cell phones. Their design was based on a novel and clever approach to assembling the reformer—including a catalytic burner and vaporizer—using layers of an inorganic ceramic material that would fire to a dense structure. The manufacturing concept was similar to Aerojet Corporation's platelet technology for making rocket nozzles and fuel cells from layers of thin stamped sheet metal diffusion bonded or brazed to seal between the layers, or to Pacific Northwest National Lab's approach to making microchannel reactors by brazing together many layers of stamped sheet metal, the stampings defining fluid flow channels in each layer. Ultimately Motorola wound down the activity.

Casio has pursued a reformed-methanol system using low-temperature PEM fuel cells targeting portable computers. The hydrogen purification method favored was palladium-alloy membrane. It is not clear how far the design and develop work progressed, or whether it is still even being pursued, and it is doubtful that a reformed-methanol PEM fuel cell system targeting laptop computers can be competitive with a direct methanol fuel cell (DMFC) system for the same application.

Although methanol is the best fuel for consumer electronics (rather than compressed hydrogen, metal hydride storage of hydrogen, or reactive metal hydrides such as sodium borohydride), the DMFC system will likely be the commercial winner in this application space for the following reasons:

DMFC systems will be simpler and less costly. It is also likely that DMFC systems will be smaller in size and weigh less than a reformed-methanol fuel cell system with equivalent electrical output. The reason is that the high operating temperature of a reformer (versus the much lower operating temperature of a DMFC system) requires that the reformer be extensively insulated and that highly efficient heat exchange be incorporated so that energy lost as waste heat is very low (and at low temperature). Insulation will increase the volumetric requirements of the reformer, and heat exchange sufficient to greatly reduce heat loss in the combustion exhaust will increase size, weight, and cost.

Toshiba is perhaps the most advanced with respect to commercial-ready product (DMFC system for a portable computer), but

Sanyo is also very active. However, barriers to commercialization will remain even after the power system is fully developed and made affordable. Perhaps the most significant barrier is a lack on consensus as to the design of methanol fuel cartridges that will safely store the fuel and interface correctly with the DMFC system. In the final analysis, will customers prefer to either carry multiple fuel cartridges or put up with the trouble and inconvenience of sourcing them locally when traveling to other cities or countries? This may be perceived as more inconvenient than carrying a couple electrical adapters in one's luggage.

7.2 Portable Power

Portable power systems are generic in function and intended to be easily carried or moved by one average adult. Typically this means that the weight must be less than 40 pounds. Furthermore the product must be of a size and shape—and with lifting handles or hand holds—that allows for one person to safely carry the product (i.e., it should not be awkward to carry and should not cause back or muscle strain). An example is the M250 product offered by Protonex Technology Corporation (see Fig. 6.2); the lifting hand holds are evident on the ends of the box.

Typically portable fuel-cell power systems are viewed as comparable to portable engine-driven generators. Common examples of low-power generators include the Honda model #EU1000IAN; All American Power model #APG3101; ETQ model #IN3500i; Honeywell model #HW1000i; Robin model R1100; and Yamaha model EF1000iS. All of these products have very similar specifications: weight is about 27 to 30 pounds, peak power is approximately 800 W to 1000 W while the continuous output rating is 750 to 900 W, and noise is moderate at about 57 to 63 decibels. With advertised selling prices in the range of US$370 to US$650,* compelling advantages of a fuel cell product with similar output power are not necessarily obvious.

* Based on author's Internet search on March 8, 2010.

Nevertheless, engine-driven generators—especially for portable applications—have three shortcomings:

- They emit toxic levels of carbon monoxide and cannot be safely operated indoors without substantial ventilation. Even when operated outdoors the carbon monoxide in the generator exhaust may be draw into air-intake ducting or poison persons working or recreating close to the generator.
- Although the noise ratings are good for engine-driven machines, 57+ decibels is still loud when people are close to the generator.
- Overall durability of generators, especially low-cost generators, is viewed as moderate to poor.

Fuel cell systems do offer fundamental advantages over all these drawbacks. To the extent that customers value low carbon monoxide emissions, quiet operation, and long operating lifetime, reformed-methanol fuel cell systems for portable power applications should gain market share. The obvious target markets for portable power systems include recreation (recreational vehicles, boating, camping, weekend cabins), construction, emergency response, and military.

SFC Energy AG has established commercial price points for portable fuel cell systems used in recreational applications mostly in Europe. Their products are based on DMFC technology and range in power from about 20 to 80 W (a 250 W product is under development). The selling price has been coming down from an initial approximate price of US$5,000 for a 20 to 30 W system to about US$3,000 presently for the same power output.

For a methanol reformer to have very low carbon monoxide emissions the reformer should be designed as depicted in Fig. 3.2 such that the fuel that is burned is a low-emission mixture of hydrogen, carbon monoxide, and carbon dioxide (this is the waste-gas stream from the hydrogen purification membrane module). Alternatively, if methanol or methanol/water mix is the fuel that is burned, then low carbon monoxide emissions may be achieved using a combustion catalyst. This will also reduce emissions of organics, including formaldehyde and formic acid, that are often present in the exhaust when methanol is the fuel.

Noise from an operating fuel cell system is easily managed by choosing balance-of-plant components carefully. The principal sources of noise are air fans and blowers, but the liquid supply pump may also be noisy. Sound can be amplified if pumps, fans, and blowers are mounted directly to panels without an intervening foam pad or spacer.

Since fuel cell systems have few moving parts, even with a methanol reformer, the promise of very good durability is real; provided pumps, fans, and blowers are designed and/or selected for long lifetime. However, like all products, durability and service intervals need to be validated through real-world testing, and unfortunately this takes time. But since fuel cells, methanol reformers, and hydrogen purifiers have all shown operational lifetime of at least several thousand hours, it is reasonable to set customer expectations for product lifetime in the range of 3,000 to 4,000 operating hours.

This is a substantial lifetime target, and one that even the best engine technology may have difficulty meeting. For example, an automobile traveling constantly at 50 miles/hour will travel 200,000 miles over 4,000 operating hours; this would represent a respectable engine lifetime even with scheduled maintenance. Low-cost generators are notorious for short lifetime on the order of 500 hours or less.

Very few companies are pursuing a commercial portable fuel cell product, perhaps because generator products offer formidable competition. IdaTech had dabbled with developing a 250 W portable product using a methanol-steam reformer and low-temperature PEM fuel cell but ultimately dropped the portable market and focused instead on stationary applications that only required about 200 to 250 W of electrical power. The product is called the iGen™; sales have been disappointing, and IdaTech has decided to no longer offer the product.

Protonex Technology Corporation is actively developing both a military and a commercial version of a portable 300 W fuel cell system also using a methanol-steam reformer and low-temperature PEM fuel cell. Though still under development, the company has sold in excess of 50 precommercial prototypes to the US Army[1] and has partnered with Cummins Onan to test and sell the recreational model once it is available.[2] If Protonex or any other manufacturer can deliver a reformed-methanol fuel cell system in the range of 120 to 300 W at a price of about US$2,000 to US$4,000, the product is expected to be successful. Ruggedized military versions could sell for up to twice that amount.

7.3 Backup Power

Backup fuel cell power systems are intended to provide electrical power to critical loads in the event that the grid fails. As such, fuel cell solutions compete against battery solutions, most commonly lead-acid batteries. Sometimes generators are used to extend the operating time of the batteries; photovoltaic panels may also be used to provide supplemental electrical power and extend the run time of the batteries. The attraction to commercialize an alternative to lead-acid storage batteries is that they simply do not last long, especially in either hot or cold climates.

Many companies are selling fuel cell systems for backup power. Examples include Altergy Systems, IdaTech, M-Field, Chung-Hsin Electric & Machinery Mfg. Corp., ReliOn, Intelligent Energy, and Dantherm Power. The vast majority of these products are operated using compressed hydrogen; however, IdaTech offers its ElectraGen™ fuel cell system in 3 and 5 kW output with optional methanol-steam reformer module. Hydrogen purification is accomplished using a palladium-alloy membrane.

For the calendar year 2008, IdaTech has reported that of the 83 ElectraGen systems sold, 78 included a methanol-steam reformer.[3] So successful have been IdaTech's reformed-methanol systems that other fuel-cell system integrators are actively seeking methanol reformers as a subsystem or component to add to their fuel cell products, displacing the need for cylinders of compressed hydrogen.

There are several reasons why backup power is an attractive application for product commercialization:

- Fuel cells offer an economic advantage over lead-acid batteries[4] and with a methanol reformer can offer much longer run time.

- The required operational lifetime is relatively low, only 500 hours to 1,000 hours in many cases, although operational reliability must be very high.
- Backup power systems are stationary and not accessible by the public.
- Methanol reformers are especially well suited for early commercialization in this field of application because size, weight, and energy efficiency are not strong drivers (contrary to most other potential applications).

The largest identified market for fuelcell backup power systems is telecommunications, both wire-line and wireless. As of 2007 there were about 4 million grid-connected telecom sites globally requiring, or benefiting from, backup power. The global growth rate is forecast to range from 5% annually to 10% annually.[5] But other critical power applications are also candidates for fuel cell systems, including border and building security. Methanol reformers are a viable alternative to compressed hydrogen because of the following reasons:

- Cylinders of compressed hydrogen are heavy and expensive; it may be cost prohibitive to supply compressed hydrogen at remote sites such as islands, mountains, and wilderness.
- It is against local fire and safety regulations to place cylinders of compressed hydrogen in some populated areas (e.g., commercial buildings).
- Liquid methanol-water mix is very easily transported (for instance in polyethylene tanks), it will not freeze, and 10 liters of methanol-water mix will yield the same amount of hydrogen as is stored in the largest commercial steel cylinders (charged to 2,400 psi).

Given that methanol reformers are in some cases competing against cylinders of compressed hydrogen, this provides a reference cost target. Cylinders of high-pressure hydrogen are often installed in groups of six to extend the run time (so-called six packs). The high-pressure manifold and housing to protect the cylinders from the elements and vandalism costs about US$1,800. One six pack would typically be installed for a fuel cell rated up to about 3 to 5 kW, larger than 5 kW, and it is likely that two six packs would be installed.

So a methanol reformer should be priced in the range of US$2,000 to $3,000 for a 2 to 3 kW fuel cell system; the benefits of the reformer allowing for some up-charge relative to compressed hydrogen. A methanol reformer scaled for a 5 kW fuel cell system may sell for US$4,000 to US$5,000. It should be expected that market penetration will suffer if product pricing is much greater than these suggested values.

Element One is a leader in offering methanol reformers to fuel cell system integrators. Their 2 kW reformer is a simplified, industrial product that is scaled to deliver 25 NLm purified product hydrogen. Simple system design and packaging contribute greatly to low cost. Other companies offering stand-alone methanol reformers include Hy9 Corporation, REB Research and Consulting, Genesis Fueltech, and Green Hydrotec. These product offerings are discussed in Chapter 6.

The future is bright for reformed-methanol fuel cell systems for backup power applications. Not only are the markets huge in size, but the present state of the art can satisfy the technical requirements for these applications. The price targets are also achievable.

7.4 Transportation

Automotive applications for combining methanol reformers with fuel cells (exclusively low-temperature PEM fuel cells) was very promising in the late 1990s and early 2000s. Many of the major global manufacturers, including Daimler Benz, General Motors, Ford Motor Corp., Toyota, Honda, Mazda, Nissan partnering with Renault, and Volkswagen,[6] were engaged in major internal development programs. However, the promise of large numbers of commercial fuel cell cars before 2010 had faded by about 2005: the technical and economic challenges are formidable.

Perhaps the most successful automotive program incorporating a methanol reformer was conducted by Daimler-Benz AG and Xcellsis, a joint venture between Daimler-Benz AG, Ballard Power Systems, and Ford Motor Corp. Following the debut of the first-generation New Electric Car (NECAR) in 1994, the effort peaked in mid-2002 with a field demonstration of NECAR 5 that consisted of a drive across the United States.[7] The trip was 3,262 miles in length and took 12 days to complete. NECAR 5, based on the Mercedes-Benz A-Class sedan, averaged a speed of 38 miles/hour (reaching a top speed of 99 miles/hour).

This fifth-generation prototype vehicle was equipped with a 15-gallon methanol fuel tank and a large battery to assist with acceleration. All of the power train (reformer, battery, fuel cell, power electronics, and balance-of-plant) was fitted beneath the passenger compartment of the sedan leaving room for four passengers.

Despite the apparent success of the Xcellsis field demonstration with NECAR 5, the US Department of Energy abandoned financial support for automotive onboard reforming, resulting in a large shift in development effort to automotive systems using compressed hydrogen stored in special high-pressure tanks.

When analyzing why the fuel cell automotive dream has not materialized, there are four reasons that consistently are cited:

- *Adequate thermal rejection from the PEM fuel cell stacks.* The fuel cell stack operates at perhaps 80°C, and this heat must be rejected in hot climates at an ambient temperature that could be as high as 45°C. The low-temperature differential combined with the large amount of thermal energy to be dissipated (as high as 50 kW) results in a rather large required radiator area. The unresolved question is, where to put this within the limited confines of a car?
- *Difficulty with performance of methanol reformers.* As attractive as it is to have a liquid fuel (methanol/water mix) that is reformed onboard the car to generate a stream of hydrogen, reformers must be hot to operate. This causes a delay of 10 minutes or more on cold startup. Methods that can be employed to overcome this problem, such as using battery power to propel the car while the reformer is heating to operating temperature, add additional cost to an already expensive power train.
- *High cost barrier to installing a hydrogen fueling infrastructure.* Rather than using an onboard methanol reformer, many automotive manufacturers have favored stored (compressed or liquefied) hydrogen onboard the car. The drawback to this approach is a severely limited refueling infrastructure and a

price tag for building such infrastructure that could easily be in the range of $1 trillion worldwide.

- *High cost of PEM fuel cell technology*. The argument has been made many times that automotive volume would drive down the cost of PEMFC stacks. But the pathway has never been explained in detail. Surely applications that can tolerate a higher price for the fuel cell stack would be commercialized first—applications including backup power, recreational power systems, and stationary power.

With time and further development effort, including perhaps resetting customer expectations for how a car should perform, the above drawbacks will almost certainly be resolved. Fuel cell propulsion systems with a reduced rated output power (perhaps only 20 to 30 kW rather than 50 to 70 kW) will help with heat rejection and overall cost reduction and take up less space. It may be that fuel cell cars are not sold in very hot climates and very cold climates. But none of these solutions will overcome the challenge of building an extensive compressed hydrogen vehicle-refueling infrastructure. Overcoming this problem is where methanol and onboard reforming excels.

General Motors has publicly stated that they intend to have a production-ready fuel cell passenger car by 2015,[8] but it is very unlikely to include a methanol reformer. Recently it was reported that Exxon Mobil Corporation was leading a consortium, including Hy9 Corporation, Ben Gurion University, Plug Power, and QuestAir Technologies, to develop an onboard reformer that could operate on ethanol, biodiesel, gasoline, and diesel fuels.[9] Plug Power has interest in applying reformers to the lift-truck market, where Plug is actively engaged with fuel cell systems operating on compressed hydrogen. If successful, the technology might also be applied to commercial automobiles; however, it should be recognized that the manner of operating an industrial forklift is much different from the way a passenger car is expected to operate. In particular, commercial forklifts operate in relatively protected indoor applications (warehouses for the most part), and it may be acceptable if a long startup time is necessary (for example, 10 or more minutes).

It is peculiar that methanol does not appear to be a fuel targeted by this development work since methanol is easier to reformer (compared with ethanol, biodiesel, gasoline, and diesel fuels), does not contain sulfur, which must be removed at significant cost from petroleum-derived fuels, and can be easily managed in an industrial application where compressed hydrogen for lift vehicles is already being accepted. Perhaps methanol is absent because Exxon Mobil does not foresee it as a mainstream transportation fuel.

Notes

1. "Protonex awarded $1.49 million contract extension by US Army," press release published on Protonex website (February 4, 2010), http://www.protonex.com/_assets/pressrelease/2d4dd9c3-8263-40dc-b0d6-6f5d4afdcbf7.pdf, accessed March 10, 2010
2. "Protonex teams with Cummins to jointly promote fuel cell power solutions in the recreational vehicle industry," press release published on Protonex website (September 15, 2009), http://www.protonex.com/_assets/pressrelease/6d30efd3-f9bf-4726-aa7f-d5b65f4c17e5.pdf, accessed March 10, 2010
3. Press release published on IdaTech website (March 30, 2009), http://www.idatech.com/uploadDocs/033009%20IdaTechPrelims.pdf, accessed March 10, 2010
4. "Early Markets: Fuel Cells for Backup Power," fact sheet published by the US Department of Energy Hydrogen Program, Energy Efficiency and Renewable Energy (November 2008)
5. S. Carlaw and C. Wheelock, "Mobile networks go green," research report published by ABI Research, Oyster Bay, New York (3Q 2008)
6. Summary table of fuel cell vehicles, created by Fuel Cells 200 and US Fuel Cell Council (October 2009), http://www.fuelcells.org/info/charts/carchart.pdf, accessed March 10, 2010
7. J. Dunne, "Fuel cells from sea to shining sea," *Popular Mech.*, 180, No. 1 (2003) pp. 79–81
8. "GM uncovers production-intent fuel cell system and reiterates 2015 commercial launch," *Fuel Cell Today*, http://www.fuelcelltoday.com/online/news/articles/2010-03/GM-Uncovers-Production-Intent-Fu, accessed on March 17, 2010
9. "Hy9 to provide membrane technology to ExxonMobil for onboard reformer project," Green Car Congress, http://www.greencarcongress.com/2008/11/hy9-to-provide.html, accessed March 17, 2010

Chapter 8

REFORMER COST—LESSONS LEARNED

As is frequently the case for new technology being developed for commercial applications, product cost* is initially a significant barrier. Reformed-methanol fuel cell systems are no different. The assumption is made that no sustainable business model can tolerate a product selling price that is equal to or less than the cost of manufacturing the product. Subsidies, tax incentives, and the like are considered short-term measures implemented to seed the market and are not intended for long-term use. The difference between cost and selling price (margin) is a variable that needs to be addressed in the context of the specific business.

Although it is easy to say that fuel cells and reformers must cost less if they are to experience widespread commercial adoption, it is far more difficult to describe a plan for significant cost reduction. This chapter will discuss some approaches that are likely to result in cost reduction without adversely affecting technical performance.

8.1 The Cost Barrier

The leading commercial applications for reformed-methanol fuel cell systems are backup power and portable power. For these

* In this chapter the term "cost" means the cost of manufacturing a product, and the term "price" means the selling price of the product.

Methanol Fuel Cell Systems: Advancing Towards Commercialization
By Dave Edlund

applications, the power range is roughly 100 W to 10 kW, but the performance specifications differ greatly between these applications (see Chapter 7). For backup power the widely accepted cost threshold varies with system output electrical power, as shown in Fig. 8.1. The non-zero intercept is common; the cost of a fuel cell system does not approach zero as the system is scaled down. This same trend is found for most, if not all, products.

The magnitude of the non-zero intercept—in this case about US$2,000 to US$2,500—is an important value for marketing purposes since it is an indication of the minimum cost of a product; any application that is not willing to accommodate this price should not be given serious consideration. Design improvements resulting in significant product cost reduction would yield a different line that is lower on the y axis, but still showing a non-zero intercept and perhaps a different slope. Also, the plot may be curved rather than linear.

For portable applications, a similar plot of selling price versus output can be constructed; however, the value of the y intercept will be less (see Fig. 8.2), and overall the line should be shifted lower than the line shown in Fig. 8.1. This fact reflects both the reduced electrical output of portable fuel cell systems (generally considered to be less than 1 kW) and also a lower acceptable price point. Chapter 7 discussed the pricing of portable system in comparison to generators, and the point was made that reformed-methanol fuel

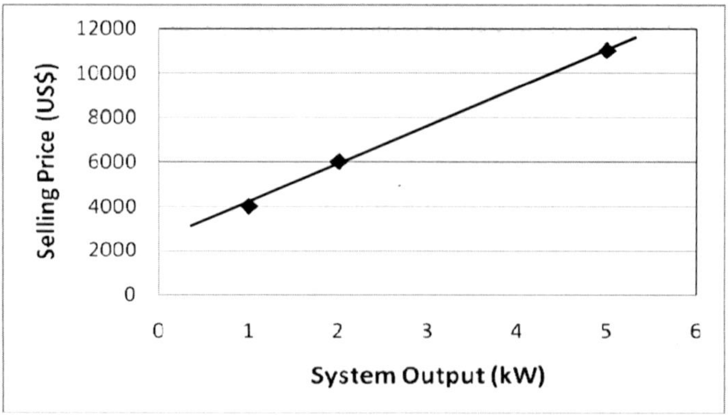

Figure 8.1 Scaling of selling price versus system electrical output for a backup-power application.

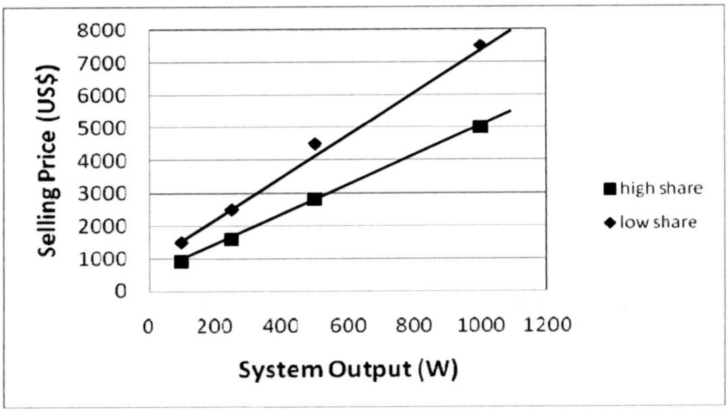

Figure 8.2 Scaling of selling price versus system electrical output for a portable power application showing the relative pricing relationship between capturing low market share and high market share.

cell systems for portable applications will have a difficult time competing on price alone. Yet, even priced as a premium product (this is the approach taken by SFC Energy AG), it is likely that fuel cell systems for portable applications will have to be aggressively priced to capture a large share of the potential market.

The curves in Figs. 8.1 and 8.2 serve as example cost barriers for the respective products. Any development effort should early on develop specific curves for their product and intended application, and if the product is a single size rather than a range of output, then the curve becomes a single point. Once the cost barrier is identified, then engineering, design, and manufacturing efforts can be productively directed at developing products that come in at a cost below the barrier.

8.2 Designing for Low Cost

Chapter 6 provides a general discussion of the cost breakdown for a reformed-methanol fuel cell system by major subsystem: fuel cell, methanol reformer, balance-of-plant, and power electronics. Now we will address the challenge of designing for low cost, ideally minimal cost since market penetration and high margins on sale of product favor low cost of manufacture.

Three factors dominate the exercise of designing for low cost: design simplicity, low-cost components, and volume. They are listed in this order because it should not be assumed that high volume is the answer to low product cost. Certainly volume helps, but alone it is not sufficient.

Using a methanol reformer subsystem for example, designing for simplicity must involve a negotiation between customer specifications (usually voiced through the marketing department) and the system design. Any feature that the customer will not pay for should be a prime candidate for elimination from the product specs. The goal is to firmly link all product specifications with the target selling price. Once this is accomplished the system design work can begin.

Unfortunately there are no simple guidelines for simplifying system design. Some generalities include considering the entire system not just an individual component. As an example (still referring to a methanol reformer) a methanol/water bleed loop controlled by a pressure-relief valve offers systemwide simplification when compared against throttling the liquid pump up and down. For the latter option, electronic control feedback to the pump, some means of flow measurement, and overpressure protection are needed.

Another example of design simplification is using a metal-membrane module for hydrogen purification rather than pressure-swing absorption (PSA). The membrane approach is a single-stage operation that functions well over a range of pressure and elevated temperature without sacrificing product hydrogen purity. On the other hand, PSA requires tighter control over temperature, a complex valve configuration, a buffer tank, or other means to smoothen product hydrogen flow since PSA is a batch process, and breakthrough detection means ensuring suitable hydrogen purity.

When it comes to hardware design, simplicity is usually apparent. Fewer parts are better; reducing the requirement to machine metal (especially alloy steels) is better; and welding is preferred over compression fittings. Figure 8.3 is an example of a complex methanol reformer design evidenced by the large number of parts, many of which are machined stainless steel. This reformer design uses a circular arrangement of 16 tubes that contain the reforming catalyst, positioned around a central burner and vaporizer coil.[1] A

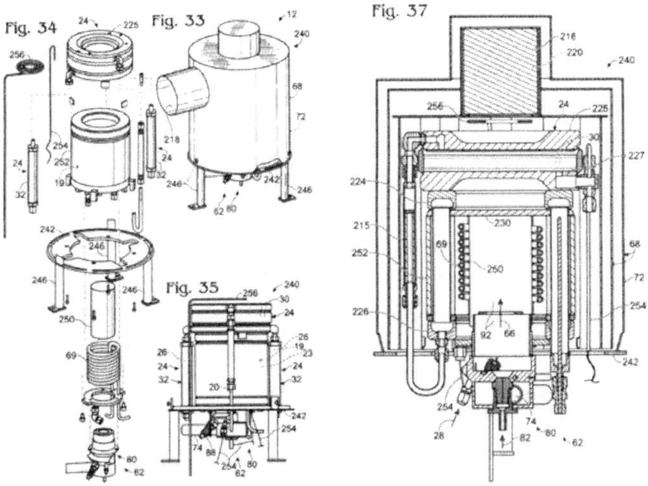

Figure 8.3 Published patent (IdaTech) illustrations of a methanol reformer using a circular arrangement of reforming tubes and a central burner: exploded view (left) and cross-sectional view (right). This is an example of a needlessly complex reformer design.

simpler design would be to use an annular reactor made from two concentric lengths of tube or pipe.

Other cost-saving design modifications would include eliminating the three legs that support the reformer in favor of a ring extending down from the reformer base (e.g., the inner tube of an annular reactor), the many compression fittings, the complex burner, and the complex base plate.

Cost scales fairly well with the number of parts for two reasons: each part brings additional indirect cost associated with ordering and inventory, and each part must be made and sold at a profit. The price that a vendor charges for a small, intrinsically inexpensive part will be increased to account for setup charges, shipping charges, and reasonable markup. Markup (or margin) is not a straight percentage that is independent of the cost of the part. For example, if a vendor were to sell 100 parts at \$500 each, a reasonable margin might be 30%, reflecting a cost of manufacture equal to \$350 (\$150 margin for each part times 100 parts equals \$15,000).

The same vendor selling 100 parts at \$2 each would likely want a larger markup than 30% (\$0.60 per piece). At 30% markup the

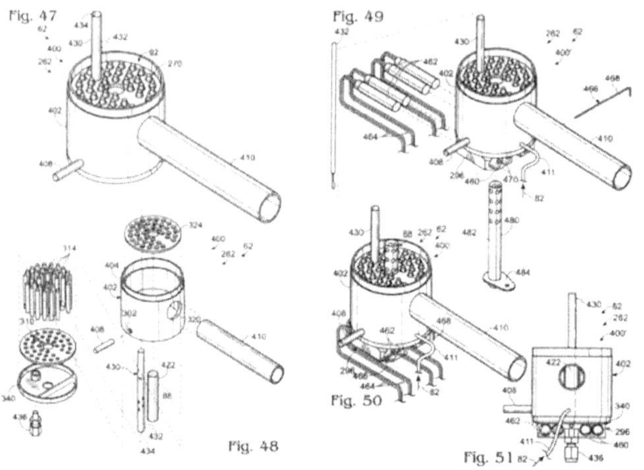

Figure 8.4 Published patent (IdaTech) illustrations of a burner assembly that is overly complex.

vendor would only receive $60 margin for the order of 100 pieces, hardly enough to cover indirect costs and make it a worthwhile order. More likely, the markup would be at least 50% (selling price of $2.8, assuming the same manufacturing cost of $1.4) or more. Thus, the more separate parts in a design the higher will be the average margin paid on the procurement of those parts, thereby increasing the cost of assembling the final product.

Another example of a complex design is the burner depicted in Fig. 8.4. The intention of this burner design is to provide a diffusion flame; that is, the fuel gas forms a flame plume as the fuel diffuses into air and combusts.[1] In this design, fuel gas is directed into a plenum beneath a collection of small-diameter tubes (Fig. 8.4, left illustration) and air is directed into the large cylindrical cavity that surrounds the collection of small tubes. The top manifold plate slips over the tubes by virtue of oversize holes in the plate, allowing air to flow out of the cylindrical cavity and exit around each of the small tubes. The illustration at the right of Fig. 8.4 shows the completed burner assembly, which includes four electric cartridge heaters to heat the base of the burner so that the liquid methanol/water fuel mix will vaporize for it to be burned as a gas.

These burner parts reflect a lot of machining, and a multitude of parts are required to assemble the burner. One reason for the complex design is that the burner is intended to function with liquid methanol/water as well as fuel gases. If the burner was only for use with gaseous fuels, a simple ribbon burner would suffice and would comprise only one part, such as a tube that is pierced with a series of holes or slots to allow escape of fuel.

Of course, many components in a reformer subsystem or fuel cell system are purchased, such as valves, thermocouples, filters, pumps, fans, and blowers. For these items the design engineer does not have the opportunity to design the component and must select from available commercial options. By understanding the minimum specifications for the complete product—including lifetime, energy efficiency, noise when operating, turndown capability, weight and physical size, emissions, operating mode (continuous or intermittent), operating environment (altitude and temperature extremes; remote site or urban), etc.—one can avoid selecting components that are more expensive than other options simply because they exceed required specifications.

8.3 Influence of Volume on Cost Reduction

As mentioned previously, increasing volume certainly helps bring down the cost. However, depending on the part or component, the cost reduction with increasing volume varies widely. Some examples are shown in Table 8.1 and are taken from actual quotations for commercial parts obtained by the author between June 2009 and March 2010. The items listed in the table were selected to represent a broad range of parts and components that are actually used in reformer subsystems and fuel cell subsystems.

Some components, notably the pressure switch, come down in cost very quickly as the volume purchased increases, whereas other components such as the hot surface igniter and the plastic fuel filter do not change much in price even as the purchase volume is increased two orders of magnitude. Figure 8.5 shows the relative cost of each component versus purchase volume. The graph is constructed from the data in Table 8.1; as the purchased volume increases, the cost is reflected as a percentage of the purchase price in quantities of 10.

Table 8.1 Influence of purchase volume on purchase price for a variety of parts and components that may be specified for use in reformed-methanol fuel cell systems (prices listed are per each item and are given in US dollars)

Sintered stainless steel filter assembly	Volume	10	50	500	2000
	Norman Filter Co.	$80.00		$68.57	$68.57
	Swagelok	$72.40	$63.71	$52.85	$51.40
Pressure switch	Volume	10	50	250	1000
	Micro Pneumatic Logic	$25.53	$9.62	$6.07	$3.87
Plastic fuel filter	Volume	10	500	2000	
	Industrial Specialties Mfg.	$8.84	$7.69	$6.32	
Reforming catalyst	Volume[1]	10	50	250	1000
	Süd-Chemie Inc.	$538	$380	$316	$285
Hot surface igniter	Volume	10	100	1000	
	Crystal Technica	$20.00	$18.00	$17.00	
Sunon 12VDC blower	Volume	10	50	250	1000
	Digi-Key	$22.04	$19.41	$17.34	$15.69
NMB 48VDC axial fan	Volume	10	50	250	1000
	Newark	$17.19	$15.01	$12.92	$12.35
Multicomp 48VDC axial fan	Newark	$19.93	$16.69	$14.58	$13.94
12VDC Solenoid valve	Volume	10	500	2000	
	Sizto Tech Corp.	$17.59	$9.22	$8.29	
Electric cartridge heater	Volume	50	500	2000	
	Tutco	$21.64	$19.47	$17.31	

[1] Volume is number of reformers, and the price is cost of catalyst for each reformer.

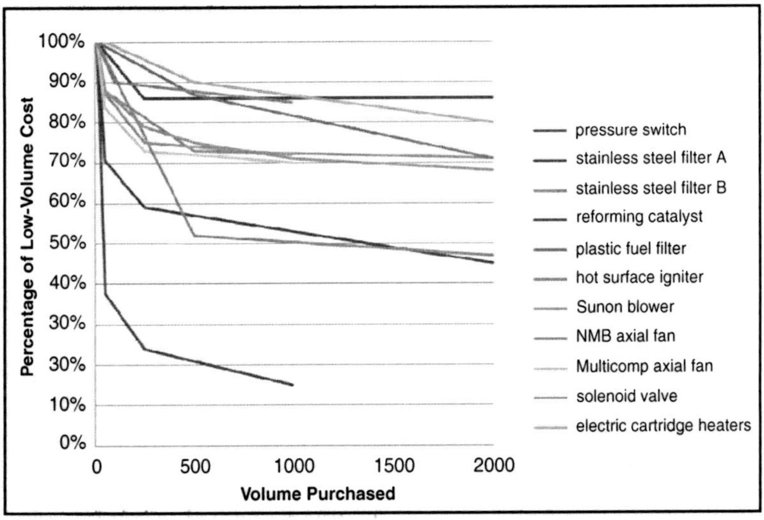

Figure 8.5 Cost reduction for the components and parts listed in Table 8.1 as a percentage of the price when purchased in quantities of 10. See also Color Insert.

Setting expectations for manufacturing cost reduction as a function of volume is important, but percentage cost reduction is widely variable from component to component. Referring to Fig. 8.5 shows that the lines fall roughly into four groupings. The uppermost grouping (including stainless steel filter A, Norman Filter Co., and the hot surface igniter) reflects that increasing purchase volume can yield small savings, the high-volume purchase price being about 85% of the purchase price in lots of 10. The next grouping shows that increasing purchase volume offers the potential of about 30% cost savings (high volume pricing is about 70% of the purchase price in lots of 10), and so on (about 50% cost savings in the next group and up to 85% cost savings for the last group, which is the pressure switch).

The danger is to assume a constant value for cost reduction as a function of manufacturing volume. As the above example illustrates, one could justifiably assume a systemwide cost savings of 15% to 50% if only a few components are quoted and the trend for those components is applied to all components and parts used in the finished product. The correct approach is to have a complete bill of materials and volume quotes for all significant items. If a quick estimation is needed, it would be prudent to assume that the cost savings on parts and components to be gained by manufacturing in higher volume is only about 20% as volume is increased from 10 systems to 1,000 systems.

8.4 Summary

Methanol reformers for fuel cell applications have been undergoing technical evolution over the past two decades, yet only recently—within the past four years—has there been a concerted effort by several companies to reduce the cost of reformers to levels necessary for commercialization to occur. Every engineer and businessperson understands that high-volume manufacturing will drive cost of production down. But the economies of volume manufacturing can only yield so much benefit. The design of the reformer, and the components specified for the reformer system, must be amenable to low-cost production.

When unique components (such as chemical reactors and hydrogen purifiers) must be designed, every effort should be exerted to make the design as simple as possible without compromising on the *required* performance features. The emphasis is on *required*, because all too often there is feature creep and added functionality finds its way into the product. More often than not, this added functionality is thought by the design engineer or the marketing product manager to be valued, when in fact the customer places little value on the added functionality and will not pay a higher price for it.

Simple designs can be recognized by fewer parts and less (rather than more) machining. Some manufacturing steps may be prohibitively expensive in small volume (e.g., forging and casting) but surprisingly affordable in high volume. Purchased components (such as valves, fittings, pumps, controllers, fans, blowers, etc.) will preferably be selected from available components used in high-volume products such as automobiles and appliances.

A decade ago, methanol reformers were only being made for demonstration programs and research, and they cost about US$20,000 to US$30,000 for a reformer with a capacity of about 20 NLm product hydrogen (adequate for a 1.5 kW low-temperature PEM fuel cell). Today, the cost is about US$6,000 for a methanol reformer with the same output capacity. But to achieve commercial success and sales of tens of thousands of reformers, the selling price must be reduced to about US$3,000 for the same reformer. The good news is that this goal is within reach.

Note

1. D.J. Edlund, D.J. Elliott, A.E. Hayes, W.A. Pledger, C. Renn, R. Stephens, and R.T. Studebaker, *Steam reforming fuel processor, burner assembly, and methods of operating the same*, Published US Patent Application 2009/0205253, August 20, 2009

Index

Color Insert

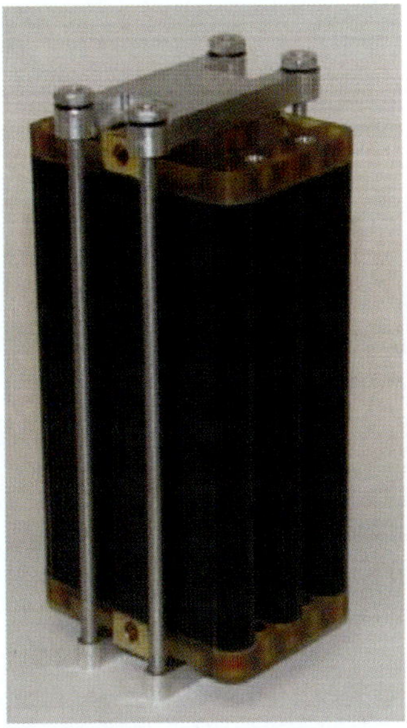

Figure 1.1 A representative PEM fuel cell stack rated to deliver a maximum of 600 W and nominal steady-state output of 310 W at 24 VDC. Fluid connections are made to the amber-colored plastic end plates. Photo courtesy of Protonex Technology Corp.

Figure 3.5 Example of an engineered catalyst support for an annular reactor. The support is a folded (corrugated) sheet metal that has been coated with catalyst (photo courtesy of Catacel Corp.).

Figure 4.2 Sample tubular palladium-alloy membrane modules sold by Power+Energy (photo courtesy of Power+Energy).

Figure 6.1 ElectraGen XTR reformed methanol fuel cell system sold by IdaTech for use as a backup power system, primarily for grid-tied telecommunications applications (photo courtesy of IdaTech, LLC).

Figure 6.2 The M250 product family under development by Protonex Technology Corporation and aimed at the recreational market (photo courtesy of Protonex Technology Corp.).

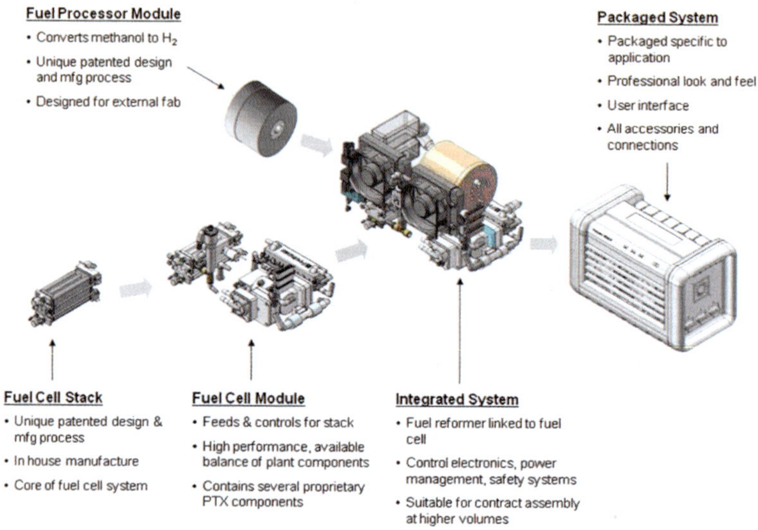

Figure 6.3 The M250 product family under development by Protonex Technology Corporation and aimed at the recreational market (figure courtesy of Protonex Technology Corp.).

Figure 6.4 A large methanol reformer rated to deliver 75 sLm product hydrogen at 99.99995% purity. Founder Robert Buxbaum is shown to the right (photo courtesy of REB Research and Consulting).

Figure 6.5 Bench-top methanol reformer rated to deliver 0.5 L/min high-purity product hydrogen (photo courtesy of REB Research and Consulting).

Figure 6.6 Model GHM-1000LPH methanol reformer with palladium hydrogen purifier. (photo courtesy of Green Hydrotec Inc.).

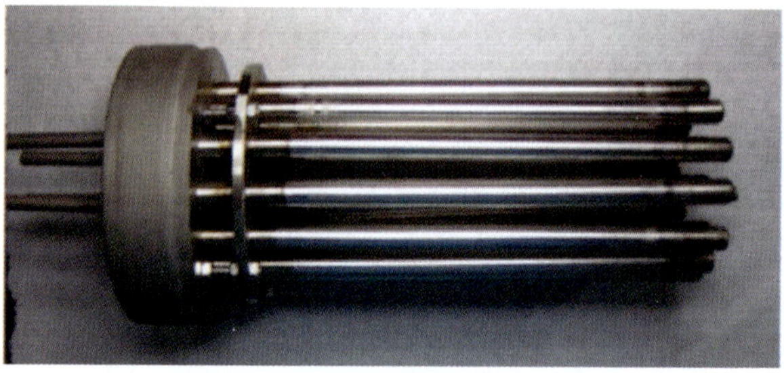

Figure 6.7 Tubular hydrogen-separation membrane bundle made by plating palladium and copper onto sintered porous stainless steel tubes. (photo courtesy of Green Hydrotec Inc.).

Figure 7.1 Angstrom fuel cell power supply (center and right) that fits within the battery space of an unmodified Motorola cell phone (photo courtesy of Angstrom Power).

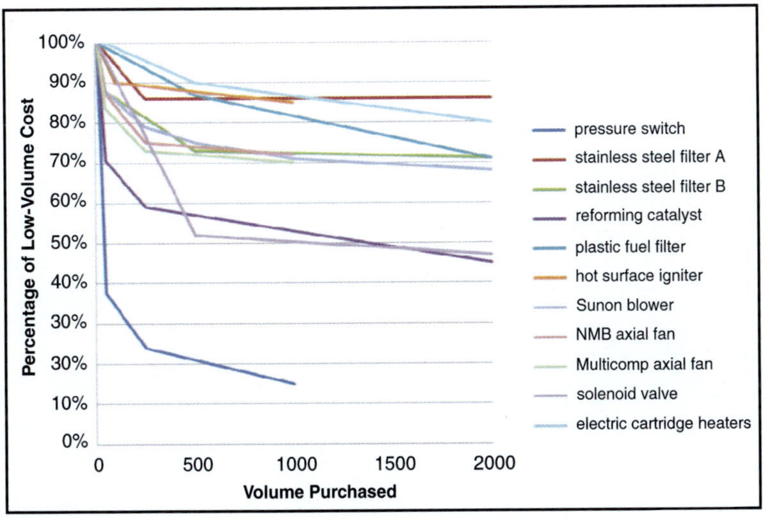

Figure 8.5 Cost reduction for the components and parts listed in Table 8.1 as a percentage of the price when purchased in quantities of 10.